An Introduction to

The Interpretation of Quantal Responses in Biology

P. S. Hewlett

Honorary Lecturer, Department of Zoology
and Applied Entomology, Imperial College, London
Detached Worker, Agricultural Research Council,
Imperial College Field Station, Ascot
and

R. L. Plackett

Professor of Statistics, School of Mathematics,
University of Newcastle upon Tyne

University Park Press
Baltimore

First published 1979 by Edward Arnold (Publishers) Limited, London
First published in the USA in 1979 by University Park Press,
233 East Redwood Street, Baltimore, Maryland 21202

Library of Congress Cataloging in Publication Data
Hewlett, Paul Soames.
An introduction to the interpretation of quantal responses in biology.
Bibliography: p.
Includes index.
1. Drugs—Dose-response relationship. 2. Probits. 3. Logits. 4. Pesticides—Dose-response relationship. 5. Biology—Mathematical models. I. Plackett, R. L., joint author. II. Title. III. Title: The interpretation of quantal responses in biology. IV. Title: Quantal responses in biology. [DNLM: 1. Quantum theory. 2. Dose-response relationship, Drug. 3. Biology. 4. Models, Theoretical. QV38.3 H612i]
Rm301.8.H48 632′.95 78-21192

ISBN 0-8391-1386-2

Printed in Great Britain

Preface

Quantal, that is 'all-or-none', responses are important in a number of branches of biology. Pharmacologists frequently use them in bioassay of drugs, toxicologists in testing the mammalian toxicities of chemical compounds. Moreover, responses of this kind are the observed effects wherever the degree of control of a pest is measured by the percentage of pest organisms killed. Thus, quantal responses are the concern of, for example, workers on fungicides, herbicides, insecticides, nematicides and molluscicides.

In this book we give a general introduction to the study of quantal responses, especially with biologists in mind. Although arising in biology, the study of quantal responses inevitably becomes somewhat mathematical. This is because a percentage response in a group of organisms can also be regarded as estimating the probability of response of each organism. Hence the theory of probability is immediately involved. However, we aim to simplify the treatment of our subject by saying relatively little about questions of estimation. The reader will find most of these dealt with in *Probit Analysis* by D. J. Finney (3rd Edition, 1971). The computations of probit analysis are commonly done on electronic computers.

We deal here with certain topics not touched on (or scarcely so) in other books, namely the dose-response curves for heterogeneous populations, the theory of monitoring of insect populations for resistance, time and response, and the general nature of quantal responses. Our treatment of quantal responses to mixtures of drugs is a wide one.

We devote two chapters to logit analysis, of which one advantage in the present field is computational. The logit transformation is, of course, almost identical in effect with the probit transformation over a very wide range of response. Like probit analysis, logit analysis presents no difficulty for electronic computers. However, the mathematics of the logistic curve is so straightforward that full logit analysis of dose-response data is easily done with far less expensive apparatus, namely with modern portable programmable electric calculators.

We are grateful to the Literary Executor of the late Sir Ronald A. Fisher, F.R.S., to Dr. Frank Yates, F.R.S., and to Longman Group Ltd., London, for permission to reprint Table IX from their book *Statistical Tables for Biological, Agricultural and Medical Research* (6th edition, 1974). We are indebted to Dr. J. Berkson and the American Statistical Association for permission to reproduce Table 4.1 from the Journal of the American Statistical Association (1953), Vol 48, pp 568–72; to Professor D. J. Finney and Cambridge University Press for permission to reproduce Fig. 3.5 from

Probit Analysis and to base Table 5.1 on values from Table II of the same work; and to the American Pharmaceutical Association for permission to reproduce Figs 2.4 and 2.5 from the paper by Dr. C. I. Bliss in the Journal of the American Pharmaceutical Association, (1944), Vol 33, pp 325–45.

We are very grateful to Mr. C. E. Dyte, Mr. C. J. Lloyd, Mr. A. R. Ludlow and Mr. T. Prickett for discussions and supply of data. Finally we express many thanks for excellent typing to Mrs. D. Sherwood, Miss P. Hunt and Miss C. A. Reynolds.

Ascot, 1978 P. S. H.
Newcastle, 1978 R. L. P.

Contents

1 Introduction

1.1 The concept of quantal response

When an individual organism is dosed with a drug the effects observed can be of two different types, graded or quantal. If the response is graded a quantitative result is observed on a single organism, such as a change in weight, a change in blood pressure, a change in the concentration of some metabolite, or a measured contraction of a muscle. On the other hand, quantal responses, otherwise called all-or-none or binary responses, are commonly observed. If the response is quantal, the organism is classified at a given time after dosage as having responded or not; in the individual organism the quantal response is a qualitative phenomenon. According to the experimental situation, the organism may, for example, be classified as asleep or awake, in oestrus or not, cured or not of some deficiency disease, paralysed or not, in convulsions or not, or as dead or alive, or, if the organism is a seed or spore, as germinated or not. In experiments relevant to pest control, classification into dead or alive will in general be most useful – in experiments, for example, with insecticides, nematicides, fungicides and herbicides.

In order to obtain repeatable and scientifically interpretable results a quantal response needs to be as precisely defined as possible. An observer needs to be able to diagnose objectively whether a response has occurred, so that he can make uniform diagnoses upon different occasions and agree with other observers. In work with insecticides, for example, an insect may often be diagnosed as 'dead' if it neither moves spontaneously nor reflexly when touched with a probe. Suitable technique can be helpful. Thus a sloping floor to a cage may assist in deciding whether a rat is asleep, since if asleep it will fail to cling to the sloping floor.

A quantal response can usually be employed to obtain quantitative results, because the per cent responding in randomly chosen groups of a given type of organism in general increases with increasing dose. The random choice is most important. The experimenter must be on his guard against unconscious non-random selection of organisms in assembling batches, and sometimes a table of random numbers is of assistance here. Another precaution is equally important, namely to ensure that the responses of the different members of a batch of organisms are independent of one another. Dependence may arise, for example, if one individual produces a chemical substance that affects the responses of others to the drug. To avoid dependence, members of a batch should not be crowded together; separate confinement of each individual may indeed be ideal, but is rarely practicable. If members of a batch do not respond independently of one another, statistical analysis of results along standard lines is invalidated.

1.2 Correction for control response

A problem that frequently arises is how to correct the per cent quantal response in a treated batch for response in an untreated control batch. Thus if a proportion of P' of organisms responds in a treated batch and a proportion C in the control, what proportion P can be regarded as responding to the drug alone? The answer normally given assumes that non-response to the drug and to the control influences are physiologically and statistically independent, enabling us to multiply the relevant probabilities. We then obtain

$$1 - P' = (1 - P)(1 - C)$$

whence, rearranging,

$$P = (P' - C)/(1 - C) \tag{1.1}$$

This expression is known as Abbott's formula.

Table 1.1 Effect of control response.

Control response %	Total response %				
	20	40	60	80	100
5	15.8	36.8	57.9	78.9	100
10	11.1	33.3	55.6	77.8	100
15	5.9	29.4	52.9	76.5	100

Table 1.1 illustrates briefly the effect of the correction for control response. We note that P is greater than $(P' - C)$.

If C is small, as it commonly is, the strict validity of equation (*1.1*) is not of much consequence, but it should be realized that it is rarely possible to test the assumptions on which the equation is based. In fact where they have been tested they have been found invalid. Kuenen *et al* (1957) and Kuenen (1957) noted that mortality of insects due to DDT stabilized, while control mortality was still rising. Moreover, Hewlett (1974) found that mortality in starved controls failed to correct mortality in starved pyrethrin-treated beetles down to the levels observed in fed pyrethrin-treated beetles. Evidently control response should be avoided if possible.

1.3 Uniformity trials for testing techniques

A technique for determining quantal responses may appear to be experimentally sound, but we may wish to test whether it is. One way of doing so is to dose each organism in a series of batches in the same way, and to determine the percentages responding in the different batches. If the technique is sound, these percentages should not be more variable than is expected from errors of random sampling. To find out, we do a chi-square test for uniformity, which is now illustrated by a numerical example. Suppose that the numbers responding/number treated in four batches are respectively 14/50, 16/50, 28/50, and 17/50. The totals are 75/200. We then calculate*

$$X^2 = \frac{\frac{14^2}{50}+\frac{16^2}{50}+\frac{28^2}{50}+\frac{17^2}{50}-\frac{75^2}{200}}{\frac{75}{200}\cdot\frac{125}{200}}$$

$$= 10.13.$$

We refer the value of 10.13 to a table of chi-square with 3 degrees of freedom (one less than the number of batches) and find $P = 0.02$. We are thus left in considerable doubt as to whether the variation in response is due to differences in random sampling, and the technique would be re-examined to see whether it could be improved, or at least a further experimental test for uniformity would be carried out. Control response does not enter the test of uniformity, since we are interested in the operation of the technique as a whole.

If a further test for uniformity was carried out, a different dose level could be used. Let us suppose that a higher level was used, and the numbers responding/number treated were 36/40, 28/40, 34/40, 34/40, 30/39, and 31/40. Calculating in a similar way to that above we find $X^2 = 6.74$ with 5 degrees of freedom whence $P = 0.2$. The two values of X^2 would be added giving a total X^2 of 16.87 with 8 degrees of freedom, whence we get an overall value of P of 0.03. We would then have to conclude that the uniformity trials as a whole threw doubt on the validity of the technique.

This procedure described above suffices for tests of uniformity when there are three or more batches. When there are only two the procedure needs to be modified. However, a test for uniformity with two batches is in fact the same as testing for a difference in the responses in two batches given different treatments, under which heading the next section considers the matter.

1.4 Comparison of two percentage responses

If we are using an experimental technique that normally gives homogeneous results (as indicated by uniformity trials), and sample sizes are substantial, a chi-square test for a 2×2 table will often be appropriate in deciding whether two treatments produce significantly different responses. As an example,

* In many books the quantity X^2 is called χ^2, but in modern practice the symbol X^2 is more usual.

suppose that two treatments produce numbers responding/number treated of 40/63 = 63.5% and 47/59 = 79.7%. We can set out the numbers as in Table 1.2.

Table 1.2 A 2 × 2 table for calculation of X^2.

Treatment	Responding	Not responding	Totals
A	40	23	63
B	47	12	59
Totals	87	35	122

We then calculate

$$X^2 = \frac{122(40 \times 12 - 23 \times 47 - \frac{1}{2} \times 122)^2}{63 \times 59 \times 35 \times 87}$$

$$= 3.14.$$

Referring to a table of chi-square, for $X^2 = 3.14$ with 1 degree of freedom, we find $P = 0.08$, so that the responses due to the two treatments are not significantly different at the 5% level. In passing it may be noted that if the response to A had been 38/63 (instead of 40/63) and the response to B had been the same as before, then $X^2 = 4.52$ and $P = 0.03$, showing significance. Thus a difference of two in the number counted as responding to treatment A could have influenced appreciably an experimenter's attitude to the results – an illustration, perhaps, of the advisability of using an objective method for diagnosing response (see § 1.1).

The fraction displayed above includes a subtraction of ($\frac{1}{2} \times 122$) in the numerator, known as a continuity correction. This is appropriate for the test with a 2 × 2 table, but is not considered worth-while when the degrees of freedom exceed one.

A chi-square test such as that just described is inadvisable if the basic (i.e. non-total) frequencies tend to be small, and the condition usually made is that none of the 'expectations' should be less than 5. An expectation may here be described as a theoretical frequency based on marginal totals. For Table 1.2 the expectation for responses to A is 87/122 of 63, or 44.9, and proceeding in this way gives the set of expectations as follows:

Treatment	Responding	Not responding
A	44.9	18.1
B	42.1	16.9

Textbooks of statistics such as Bailey (1959), Snedecor and Cochran (1967), and Goodman (1970), discuss chi-square tests more fully. They also describe methods for testing significance where frequencies are small. Pearson and Hartley (1969) and Mainland, Herrera and Sutcliffe (1956) give special tables facilitating rapid tests of significance for 2×2 tables.

Analysis of heterogeneous data is not included in this book.

2 The typical dose-response curve

2.1 The shape of the typical curve

Suppose that a series of large, randomly chosen groups of organisms is taken, and that doses of drug extending over a range are given respectively to the different groups; then, when per cent responses (corrected for control) are plotted against dose, the plotted points usually fall near to a curve of a fairly simple kind. The curve is usually an asymmetric sigmoid (Fig. 2.1) such that the left-hand, lower limb of the curve is shorter than the upper, right-hand one, so that the point of inflexion of the curve is lower than the 50% level.

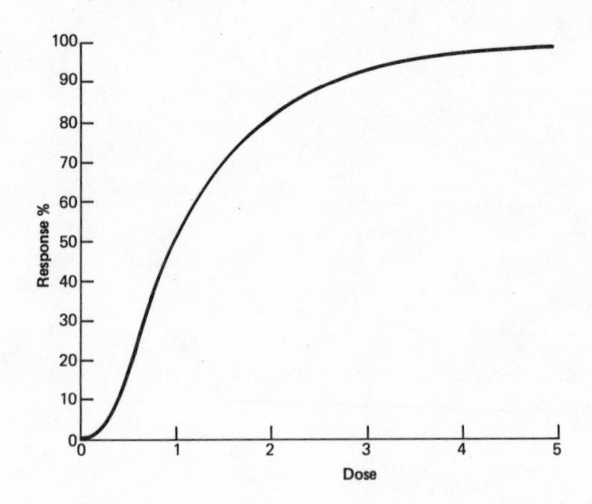

Fig. 2.1 A typical asymmetric sigmoid curve relating per cent response to dose.

From this curve we can estimate the various doses producing the corresponding per cent responses, and the dose producing K% response is known as the effective dose K or EDK. Thus the doses producing 30, 50, 80 or 90% response are known as the ED30, ED50, ED80 and ED90, and so on. We notice, however, that the upper limb of the curve is asymptotic to the 100% level, so that it is impracticable to read off the ED100, which is indeterminate. Where death is the response we can read off the lethal dose K, and speak of the LD30, LD50, and so on.

In the literature one sometimes sees claims that an ED100 or LD100 has been determined, but such claims are unsound scientifically.

2.2 The effect of taking logarithms

If instead of plotting per cent response against dose we plot per cent response against log-dose, we usually arrive at a symmetric sigmoid curve in place of the asymmetric (Fig. 2.2). The effect of taking logarithms can be pictured as a

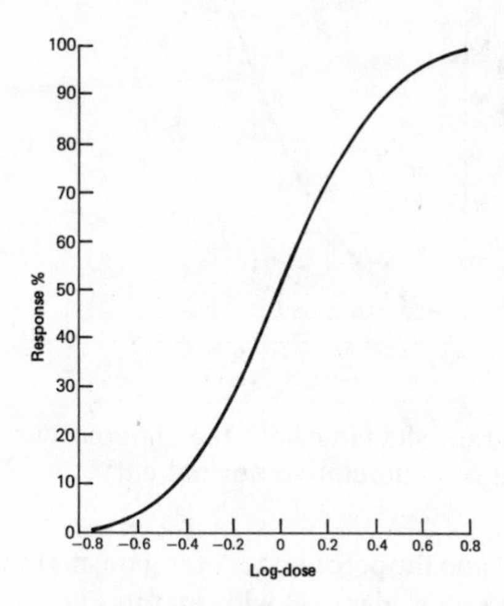

Fig. 2.2 A typical symmetric sigmoid curve relating per cent response to log-dose.

progressive compression of the abscissa-axis, the compression becoming the greater as we proceed to the right. The resultant sigmoid has its point of inflexion at the 50% point.

2.3 Interpretation of the log-dose-response curve

This curve can be interpreted in terms of what are known as individual tolerances. The individual tolerance is a hypothetical quantity characteristic of the individual organism with respect to the particular drug used. The tolerance is the dose of drug of such a magnitude as to be just insufficient to make the individual organism concerned show the quantal response. Thus if a dose K_1 causes 65% of organisms to respond, 65% of them have tolerances of up to K_1. Similarly if a dose K_2 causes 55% of organisms to respond, 55% of them have tolerances of up to K_2. Hence $10\% = (65-55)\%$ of the organisms have tolerances between K_1 and K_2 of the drug. This argument is evidently a general one.

We can refer to Fig. 2.2 and find the proportion of tolerances falling within any range of log-amounts of drug. In Fig. 2.3, equal successive ranges of log-

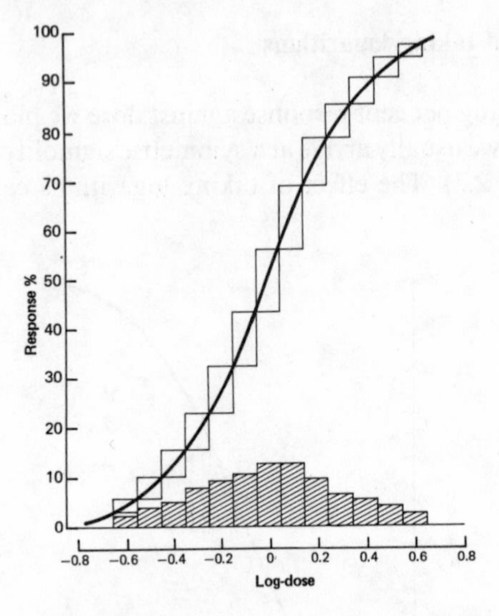

Fig. 2.3 A step diagram showing how the sigmoid curve relating per cent response to log-dose is a cumulative normal curve.

dose have been taken and the percentages of organisms have been indicated on the sigmoid curve, which is identical with that in Fig. 2.2. The heights of the steps represent the percentages of tolerances falling within the successive ranges of log-dose. The step-blocks have then been brought down to the base line in order to show the statistical distribution of log-tolerances.

The log-tolerances have a symmetric distribution which is usually close to normal. We may thus state that the symmetric sigmoid curve relating per cent response to log-dose is a cumulative normal distribution of log-tolerances. Alternatively we may state that the asymmetric sigmoid curve relating per cent mortality to dose is a cumulative lognormal distribution of tolerances.

We note that the symmetric sigmoid curve relating per cent response to log-dose can vary in position, i.e. in the value of the log-ED50, and also in the sideways extension of the curve, i.e. the curve can be more steep or less so. It will vary in these respects with the drug, the organism and the type of response. Variation in position implies variation in the mean log-tolerance (and hence in the median effective dose), while the variation in steepness implies variation in the standard deviation of the log-tolerances, the larger the standard deviation the more spread the curve.

2.4 Mathematical statement

So far it has been shown by informal, graphical methods that the typical dose-response curve arises as a cumulative distribution of tolerances. We may now present this situation in mathematical terms. If z is the dose, the distribution of tolerances may be given as

$$dP = f(z)\,dz \tag{2.1}$$

This states that the proportion, dP, of tolerances lies between z and $z+dz$, where dz is a small interval on the dose scale. Moreover, $f(z)$ is the frequency function relating dP to dz and is uniquely determined for each possible value of z. Thus if a dose z_0 were given, by (*2.1*),

$$P = \int_0^{z_0} f(z)\,dz \tag{2.2}$$

and if the dose were infinite we would expect

$$P = \int_0^{\infty} f(z)\,dz = 1 \tag{2.3}$$

We have stated that the distribution of log-tolerances is usually normal, so we can be more specific and formulate dP as

$$dP = \frac{1}{\sigma(2\pi)^{\frac{1}{2}}} \exp\left\{-\frac{(\log z-\mu)^2}{2\sigma^2}\right\} d(\log z) \tag{2.4}$$

The quantities μ and σ are respectively the mean and standard deviation, namely the parameters, of the normal distribution concerned.

2.5 Experimental support for the concept of tolerance

Though plausible, the interpretation of the dose-response curve as a cumulative distribution of tolerances cannot be regarded as wholly satisfactory, since the evidence so far given for tolerances is indirect. There are, however, certain special circumstances in which tolerances, or something very like them, can be observed directly. This occurs in the assay of digitalis on cats.

The technique consists of anaesthetizing a cat and then injecting a digitalis preparation steadily and slowly, or by frequently repeated increments, until the cat's heart ceases to beat. The total amount injected is then a measure of the lethal dose peculiar to the individual cat concerned.

Bliss (1944) gave an account of large-scale collaborative experiments by different laboratories with assays of this type. After various necessary corrections, Bliss gave the distribution of lethal doses shown in Fig. 2.4. The distribution is significantly skewed to the right, Bliss giving $g_1 = 0.336 \pm 0.166$ as a test for skewness. Bliss also gave the distribution of the log-lethal-doses shown in Fig. 2.5, in which the distribution is approximately normal. Bliss gave no test for normality, but a chi-square test gives $X^2 = 10.6$ with 13

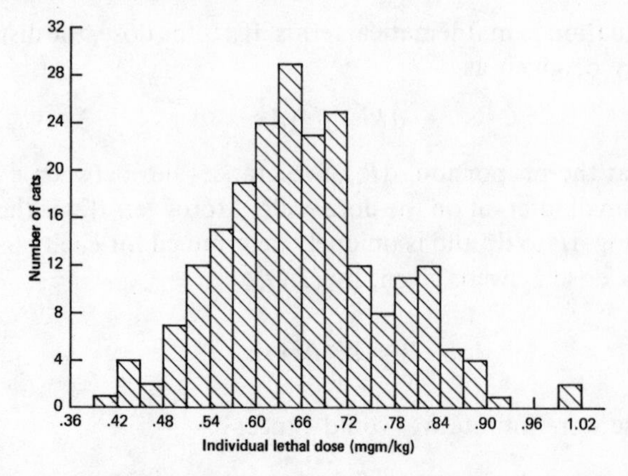

Fig. 2.4 Frequency distribution of the individual lethal doses obtained with a tincture of digitalis with 216 cats (after Bliss, 1944).

degrees of freedom, whence $P = 0.64$; the frequencies for the log-doses do not contradict a hypothesis of normality.

These experiments, and similar ones with digitalis, lend support to the concept of a lognormal distribution of tolerances. Certain caveats must be entered however, since, in particular, the rate of injection of cardiac glycosides is known to influence the results. Moreoever, there may be a time lag between injection of digitalis and its taking effect, and slow infusion need not have quantitatively the same effect as virtually instantaneous dosage by a single injection.

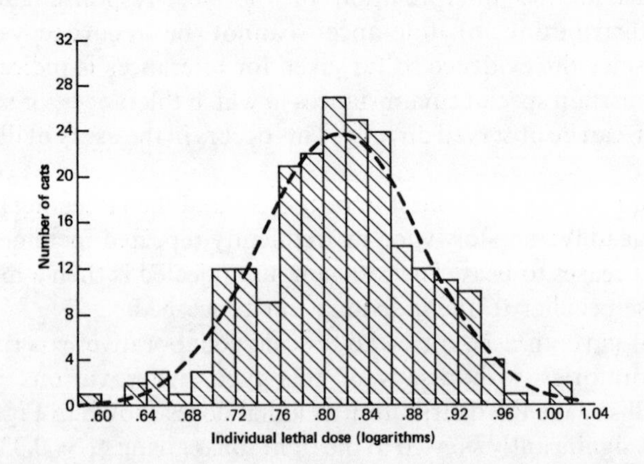

Fig. 2.5 Frequency distribution on an equal logarithmic scale of the data shown in Fig. 2.4. The smooth curve is that expected from the theoretical distribution.

2.6 Atypical response curves

Certain departures from the response curve of typical shape do sometimes occur, especially owing to the presence of a minority of unusually resistant or unusually susceptible individuals. This topic is dealt with in a later chapter, but some grossly atypical response curves may be mentioned here. These are discussed more fully by Finney (1971, p. 265). The curves concern certain fungicides. What happens is that over a low range of dose mortality of the spores increases with increasing dose. When, however, the dose is increased further, mortality of the spores actually falls, before rising again eventually. The relation between response and dose is not monotonic. The probable explanation is that in effect two fungicides are operating, one in the dissociated form and the other in the non-dissociated, the former at lower concentrations and the latter at higher. Finney shows that such a hypothesis accounts at least semi-quantitatively for the very unusual and seemingly anomalous results.

3 The Probit and Similar Transformations

3.1 Representation of linear relations by adjacent scales

The most usual method of representing graphically a linear relation is by means of the cartesian plot, but there is another method wherein the relation is represented by uniform linear scales placed next to one another. This can be easily illustrated by the relation between temperature Fahrenheit and temperature Celsius. As an equation we have

$$F = 32 + (9/5)C,$$

shown as a cartesian plot in Fig. 3.1. We can now transfer the axis markers on to the straight line representing the relationship, and turn the axis markers at right angles to the line (Fig. 3.2). Thus corresponding temperatures on the two scales can be converted merely by reading the scales on either side of the line.

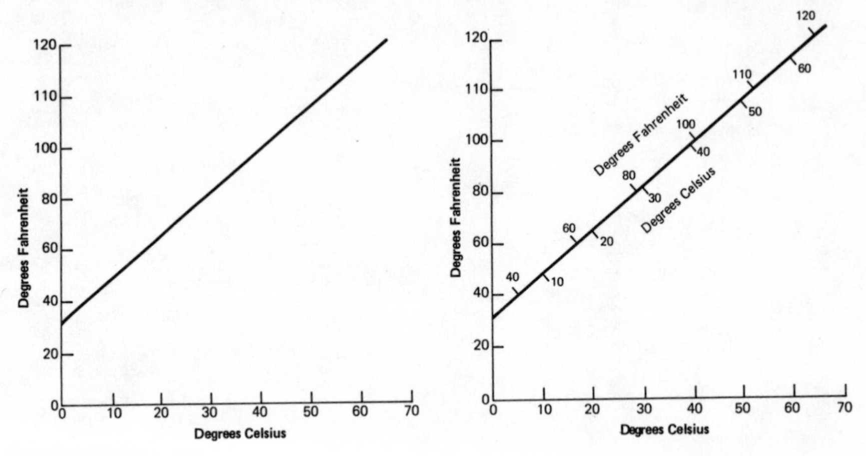

Fig. 3.1 A cartesian plot of degrees Fahrenheit against degrees Celsius.

Fig. 3.2 A plot illustrating how the relation between degrees Fahrenheit and degrees Celsius can be represented as two adjacent uniform scales.

These two scales are known as uniform scales because each unit is represented by the same distance throughout each scale (compare, say, a log-scale). The argument is evidently general; any linear relation can be represented by two adjacent uniform scales, while two adjacent uniform scales represent a linear relation. The relevance of this consideration to the typical dose-response curve will become apparent shortly.

3.2 An informal explanation of the probit transformation

We now return to the typical curve-relating per cent response to log-dose. Fig. 3.3 shows this once again. However, a set of equally spaced perpendiculars has been drawn from the curve to the axis of log-dose. In principle, these perpendiculars could have been arbitrarily but evenly spaced along the abscissa axis, but it is sensible to space them in terms of standard deviations of the cumulative normal curve concerned. The scale so defined is known as that of normal equivalent deviates (N.E.D.) or normits. Evidently the scale of N.E.D. and of log-dose are two uniform scales, so that the N.E.D. are linear in log-dose. Moreover, the intercepts of the N.E.D. on the percentage axis, read as N.E.D., will evidently be linear in log-dose. Thus, by measuring the percentage response in terms of standard deviations we arrive at a linear relation between the N.E.D. and log-dose.

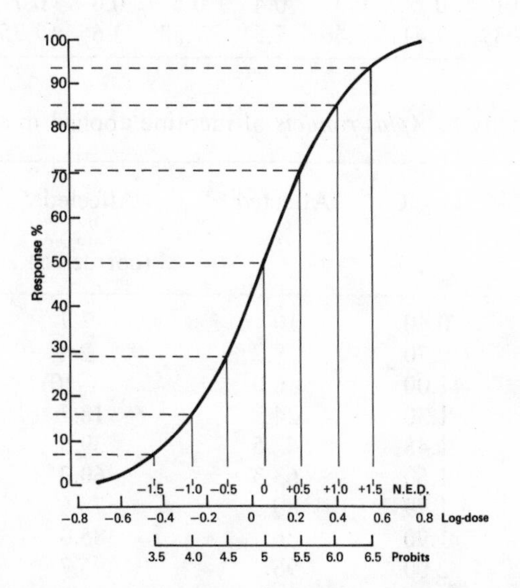

Fig. 3.3 A graphical explanation of the probit transformation.

There is, however, one further modification to be made. Biologists have generally preferred to use measures of response that are all positive. This is achieved by adding five to the N.E.D., and calling the resultant measures of response probits (an abbreviation for probability units). Probits, of course, remain linear in log-dose. Table 3.1 gives a short table for the transformation of percentage to probits. Finney (1971) gives fuller tables. By plotting probit mortality against log-dose for empirical data we usually obtain an approximately linear plot.

Colquhoun's (1971) explanation of the probit transformation is in some ways similar to ours, but is more elaborate.

Table 3.1 Transformation of percentages to probits.

%	0	1	2	3	4	5	6	7	8	9
0	—	2.67	2.95	3.12	3.25	3.36	3.45	3.52	3.59	3.66
10	3.72	3.77	3.82	3.87	3.92	3.96	4.01	4.05	4.08	4.12
20	4.16	4.19	4.23	4.26	4.29	4.33	4.36	4.39	4.42	4.45
30	4.48	4.50	4.53	4.56	4.59	4.61	4.64	4.67	4.69	4.72
40	4.75	4.77	4.80	4.82	4.85	4.87	4.90	4.92	4.95	4.97
50	5.00	5.03	5.05	5.08	5.10	5.13	5.15	5.18	5.20	5.23
60	5.25	5.28	5.31	5.33	5.36	5.39	5.41	5.44	5.47	5.50
70	5.52	5.55	5.58	5.61	5.64	5.67	5.71	5.74	5.77	5.81
80	5.84	5.88	5.92	5.95	5.99	6.04	6.08	6.13	6.18	6.23
90	6.28	6.34	6.41	6.48	6.55	6.64	6.75	6.88	7.05	7.33
—	0.0	0.1	0.2	0.3	0.4	0.5	0.6	0.7	0.8	0.9
99	7.33	7.37	7.41	7.46	7.51	7.58	7.65	7.75	7.88	8.09

Table 3.2 Toxicity to *Aphis rumicis* of nicotine applied in a spray.

C (= dose)	Log C	Affected %	Affected % (corrected)	Probit response
2.5	0.40	10	3.7	3.21
5.0	0.70	8.7	2.4	3.00
10	1.00	6.0	— (0)	—
20	1.30	24	18.7	4.11
30	1.48	43.5	39.6	4.74
40	1.60	63.3	60.7	5.27
60	1.79	80	78.6	5.79
80	1.90	86	85.0	6.04
100	2.00	96	95.7	6.72
150	2.18	96	95.7	6.72
200	2.30	100	100	—
Control		6.5	—	

Note: $C = 1000 \times$ conc. % of nicotine. 45–50 aphids per group.

3.3 An experimental example

Details of a relevant set of experimental results can now be given, namely from an experiment recorded by Tattersfield and Morris (1924). Groups of 45–50 aphids were sprayed with respective concentrations of nicotine, and Table 3.2 gives the percentages of aphids affected. Fig. 3.4 shows probit of per cent affected (corrected for control) plotted against log-concentration of nicotine.

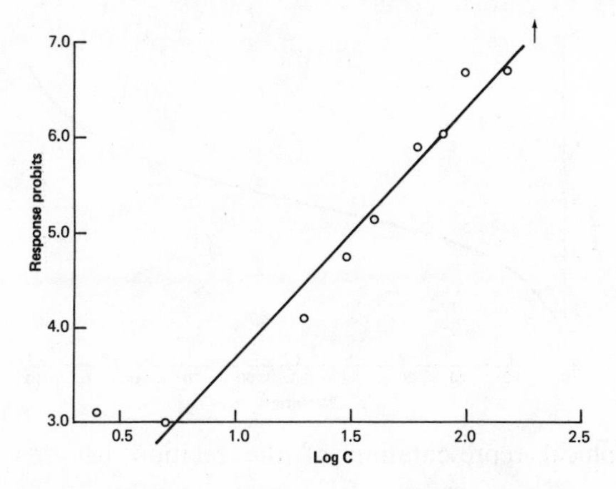

Fig. 3.4 An experimental example of a linear relation between probit response and log-dose.

The experimental points fit well to a straight line, and the reader if he desires can plot per cent mortality against dose to see the effect of taking logarithms and the plotting of probits instead of percentages.

3.4 Mathematical statement of the probit transformation

Readers sufficiently mathematically inclined will desire a mathematical account of the probit transformation. Historically the normal equivalent deviate predates the probit by a year and is slightly simpler theoretically. However, we shall present the probit transformation. The probit of the proportion P is defined as the abscissa which corresponds to a probability P in a normal distribution with mean 5 and variance 1, or symbolically

$$P = \frac{1}{\sqrt{(2\pi)}} \int_{-\infty}^{Y-5} \exp\left(-\tfrac{1}{2}u^2\right) \mathrm{d}u \qquad (3.1)$$

If (*2.4*) represents the distribution of tolerances on the $\log z = x$ scale of doses, the expected proportion of organisms responding to a dose of $\log z_0$ is

$$P = \frac{1}{\sigma\sqrt{(2\pi)}} \int_{-\infty}^{\log z_0} \exp\left[-\frac{1}{2\sigma^2}(\log z - \mu)^2\right] \mathrm{d}x \qquad (3.2)$$

Comparison of the two expressions for P in (*3.1*) and (*3.2*) then shows that the probit

$$Y = 5 + (1/\sigma)(\log z - \mu) \qquad (3.3)$$

Thus Y is linear in $\log z$, the line having a slope of $(1/\sigma)$, which it will be useful to designate β, or

$$Y = \alpha + \beta \log z. \qquad (3.4)$$

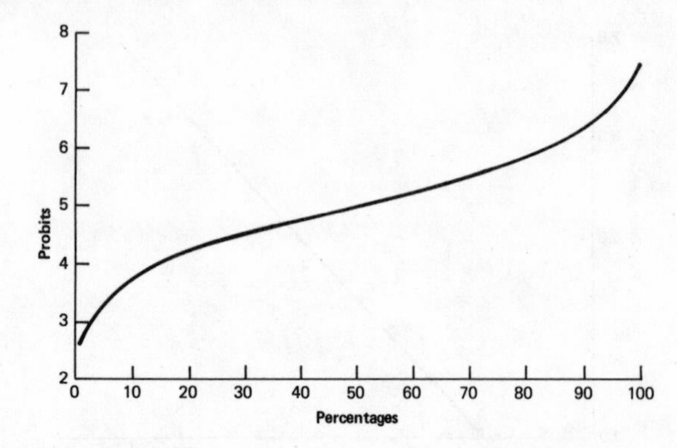

Fig. 3.5 Graphical representation of the relation between probits and percentages.

Fig. 3.5 shows the relation between percentages and probits.

For graphical analysis of experimental results, tables can be used to find the probits of responses to be plotted against log-doses, also looked up in tables. However, for the graphical analysis, special graph paper, known as log-probability paper, is available commercially. Such paper bears scales that carry out the two transformations. Those forms of the paper with scales appropriately arranged are an aid to quick analysis; other forms result in diminutive diagrams.

3.5 Relative potency and bioassay

It is often desirable and useful to be able to quote what is known as a relative potency of one biologically active preparation relative to another. This is

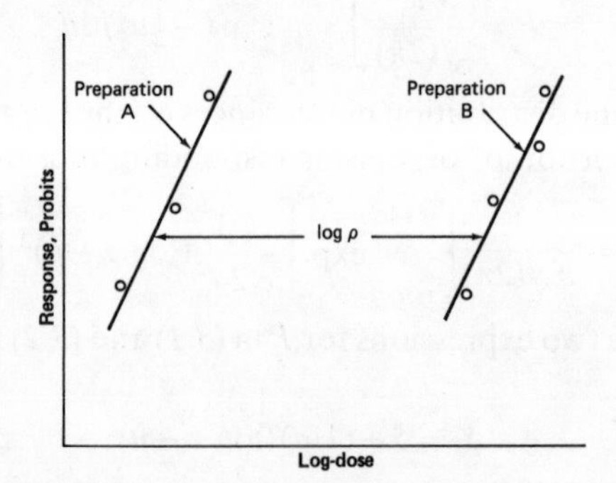

Fig. 3.6 The relative potency of two biologically active preparations giving parallel probit-log-dose lines. Preparation A is ρ times as active as preparation B.

possible, if the probit-log-dose transformation applies, only if the probit-log-dose lines for the respective preparations may be taken as parallel. The relative potency is then the antilog of the horizontal distance, between the parallel lines, measured on the log-scale (Fig. 3.6). One preparation may then be said to be so-many times as potent as the other.

If the lines are not parallel a statement of relative potency is not possible, though if the lines are well separated a statement of the ratio of the ED50's may be of some use in practice.

3.6 Other transformations

At various times the merits of transformations other than the probit have been pointed out. These include the logit and angular transformations, and in particular the cumulative logistic curve is very similar to the cumulative normal. Logits, however, have never achieved much popularity among biologists analysing dose-response data. On the other hand, statisticians increasingly see advantages in logits and we give logits substantial attention in this book (Chapters 4 and 6). Many computer programs for analysing dose-response data at present use either transformation.

The theoretical differences between normits and probits are of course trivial, though the former have slight theoretical advantages over probits in dealing with mixtures of drugs (see Chapter 7).

3.7 The probit plane

Up to now we have considered dose as a single, simple variable. Sometimes, however, dose must be considered as consisting of two components. This happens with fumigants when dose consists of concentration of fumigant and time of exposure. Often probit mortality can then be considered as a plane

$$Y = \alpha + \beta_1 \log C + \beta_2 \log T. \qquad (3.5)$$

If this relation holds, for constant response, $C^n T =$ constant where $n = \beta_1/\beta_2$. Only where $\beta_1 = \beta_2$ does $CT = 1$.

A further circumstance in which dose has two components is where an insecticidal spray is varied in two ways, one in which concentration of the toxic principle is varied and the other in which the total deposit of spray is varied. Here again a probit plane is often applicable.

$$Y = \alpha + \beta_1 \log z_1 + \beta_2 \log z_2 \qquad (3.6)$$

where z_1 and z_2 are concentration and deposit. Finney (1971) deals with applications of (*3.5*) and (*3.6*).

3.8 Incomplete dose control

An obvious feature of certain experimental situations is that the mean dose administered to a batch is well controlled, but the doses administered to the individuals of the group are less so. This happens for example when a batch is sprayed. It is obviously interesting to know what effect this has upon dose-response data, and Finney (1971, p. 196) considered this problem.

If there is complete control over dose suppose that

$$Y = \alpha + \beta \log z. \tag{3.7}$$

If the control is less than complete, suppose that the mean log-dose is ξ with variance γ^2. It can then be shown that

$$Y = 5 + (\alpha - 5 + \beta\xi)/(1 + \beta^2\gamma^2)^{\frac{1}{2}}. \tag{3.8}$$

This means that although, on average, the estimate of the log-ED50 is unaffected, the estimate of β is lowered to $\beta' = \beta/(1+\beta^2\gamma^2)^{\frac{1}{2}}$. Hewlett (1954) calculated a table of β for various values of γ. The effect was not as great as might have been expected. For example with a coefficient of variation of 20 %, $\beta = 6$ was lowered to $\beta' = 5.32$.

4 The logit transformation

4.1 An informal explanation of the logit transformation

Consider again a typical curve relating per cent response to log-dose, shown in Fig. 4.1. A set of equally spaced perpendiculars has been drawn from the curve to the axis of log-dose. In Fig. 3.3, the intercepts on the percentage axis were normal equivalent deviations or probits. However, the curve in Fig. 4.1 is

Table 4.1 Transformation of percentages to logits. The text explains how to use this table.

%	0	1	2	3	4	5	6	7	8	9
50	0	0.04	0.08	0.12	0.16	0.20	0.24	0.28	0.32	0.36
60	0.41	0.45	0.49	0.53	0.58	0.62	0.66	0.71	0.75	0.80
70	0.85	0.90	0.94	0.99	1.05	1.10	1.15	1.21	1.27	1.32
80	1.38	1.45	1.52	1.59	1.66	1.73	1.82	1.90	1.99	2.09
90	2.20	2.31	2.44	2.59	2.75	2.94	3.18	3.48	3.89	4.60
—	0.0	0.1	0.2	0.3	0.4	0.5	0.6	0.7	0.8	0.9
99	4.60	4.70	4.82	4.95	5.11	5.29	5.52	5.81	6.21	6.91

not exactly the same as in Fig. 3.3. The distribution of tolerances is now assumed to be logistic instead of normal. This means that the intercepts become logits instead of probits. A plot of logit mortality against log-dose for empirical data will usually also be approximately linear. Table 4.1 is a short

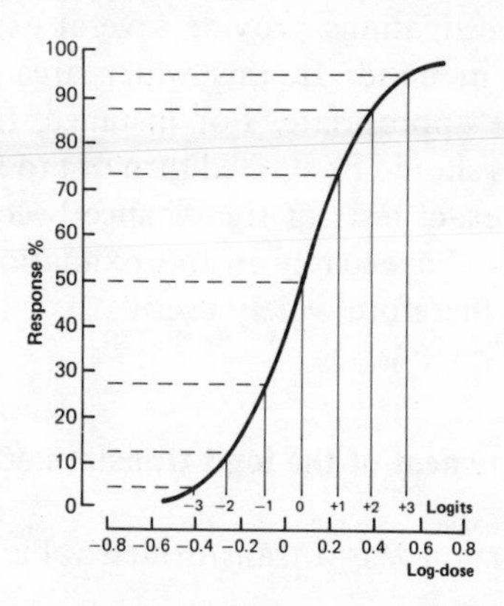

Fig. 4.1 A graphical explanation of the logit transformation.

table for the transformation of percentages to logits. Ashton (1972) gives a fuller table. Table 4.1 gives only percentages greater than 50. The logit of a percentage, p, less than 50, is minus the logit of percentage 100-p.

Example 1. Find the logit for 75 %. We read along the row beginning 70 until the column headed 5 is reached. The value required is 1.10.

Example 2. Find the logit for 25 %. Since 25 is less than 50, we look up the logit for 75 % and change the sign. The value is − 1.10.

4.2 A comparison of probits and logits

Probits and logits are based respectively on the normal and logistic tolerance distributions. The two distributions can be compared by standardising them so that each has zero mean and unit variance. Mortalities corresponding to a given log-dose can then be derived by reference to the standard scales. We find that the two tolerance distributions agree closely (within 1.3 %) for responses which range from 1–99 % (Cox, 1970). In fact, the difference between them can be detected only by experiments using extremely large samples. An immediate consequence is that statistical analyses based on the two transformations will always lead to essentially the same results.

Statistical methods using the probit transformation are well established. The question naturally arises whether logits need serious consideration, in view of their close agreement with probits when suitably scaled. Reasons which support the use of logits are as follows. In this book we deal mostly with situations in which there is a single explanatory variable, namely log-dose; here computations involving logits are simpler than those involving probits because the logit of a percentage and its weight are easier to calculate (especially with the modern portable electric calculator). Moreover, in more complex experiments quantal response may be related to measured variables like log-dose, together with categorical variables like other quantal responses. Long-term medical investigations provide several examples of the type of investigation we have in mind. In this wider area of enquiry, the logit transformation is more appropriate. For instance, the quantities that the transformation suggests should be studied turn out to have natural meanings in the context. Again, exact tests of significance become available in small samples, so that we do not have to rely on approximations like chi-square. The logit transformation is therefore widely useful.

4.3 Mathematical statement of the logit transformation

The logit of the proportion P is a transformed value l defined by

$$l = \log_e\{P/(1-P)\}. \qquad (4.1)$$

Here the suffix e means that a natural logarithm is taken. By analogy with equation (*3.4*) we assume a linear relation between the logit and log-dose, expressed by

$$l = \theta + \phi \log z \qquad (4.2)$$

Here logarithms without suffix are normally to base 10. We have introduced two further parameters θ and ϕ in order to distinguish them from α and β used previously. However, the comments in the previous section imply that each pair of parameters can be derived from the other to a very close approximation. In particular,

$$\phi = 1.814\ \beta$$

where the constant is near the correct value of $\pi/\sqrt{3}$, and, equating log-ED50's,

$$-\theta/\phi = (5-\alpha)/\beta.$$

5 Calculations in probit analysis

5.1 Maximum likelihood fitting of a probit line

If a set of data relating quantal response to dose is to be analysed, it is often sufficient to plot probits (looked up in Table 3.1) against log-dose, fit a straight line by eye, and accept graphical estimates of the parameters of the line. The parameters concerned are α and β in equation (*3.4*), of which we can make estimates, say a and b. Often, however, interest centres on finding m, an estimate of the log-ED50, together with b. Of course

$$m = (5-a)/b. \qquad (5.1)$$

Other log-ED's are sometimes needed. Obviously fitting a line by eye is somewhat subjective, and does not give the best possible estimates of the parameters, nor give any estimates of their errors.

For objective estimation of the parameters the method of maximum likelihood is normally used in probit analysis. Though not necessarily ideal (Finney, 1971), this method is greatly superior to graphical analysis, and yields estimates of the errors. However, for probit analysis the method is somewhat complex and its full implementation requires an electronic computer, so only rather general remarks need to be made here. The method of maximum likelihood fitting of a probit-log-dose line can be described as an iterative weighted regression procedure. In the simple regression procedure, which is described in elementary textbooks on statistics, a straightforward method for calculating a line is possible because each point on the plot is taken to have the same weight, i.e. to carry the same amount of information. In a probit plot, on the other hand, points carry different weights, and weights must be taken into account for calculating the line. The weight depends on n, the number of subjects in the group, and on a quantity called the weighting coefficient, w; the weight is equal to the product of the two, nw. The coefficient, w, varies according to the level of response in the group, and according to the proportional control response, C. In fact

$$w = Z^2/Q[P+C/(1-C)]. \qquad (5.2)$$

Here

$$Z = (2\pi)^{-\frac{1}{2}} \exp\left[-\tfrac{1}{2}(Y-5)^2\right], \qquad (5.3)$$

that is, the ordinate to the standardized normal frequency function for $(Y-5)$. P and Q are the respective proportions for response and non-response corresponding to Y. Table 5.1 shows values of w for different values of Y and

C. For $C = 0$, w falls symmetrically from a maximum as Y departs upward and downward from 5. For a given Y, w falls as C rises.

If for each group of subjects we could evaluate w directly from each empirical probit (found from the observed per cent response from Table 3.1), calculation of the maximum likelihood probit-log-dose line would be fairly easy; a straightforward scheme of calculation would give the line directly. But theory shows that weighting coefficients are not legitimately obtained by the procedure just supposed. Instead, the correct weighting coefficient must be found from that value of the probit, Y, read, for the log-dose concerned, from the maximum likelihood line. Thus it might appear that the maximum likelihood line can be calculated only if it is known already. However, the way out of the seeming *impasse* is a process of successive approximations. A provisional line is fitted to the points in some way, e.g. by simple regression or by eye, and provisional weights are found after determining weights on the basis of probits, read from the provisional line, for the dose levels concerned. A sequence of calculations, known as a cycle, or iteration, is then done, which gives a line closer than the provisional to the maximum likelihood line. Weighting coefficients are found from the new line and a further, similar cycle is performed. By proceeding in this manner the maximum likelihood line can be approached to any required degree of accuracy.

5.2 First-cycle computations

As already indicated, full maximum likelihood fitting of a probit-log-dose line requires an electronic computer. If such fitting is impracticable, a single cycle of computations based on an eye-fitted line will normally give a reasonable approximation to the maximum likelihood line. These computations can be done with a desk computer. Finney (1971) describes them, giving such details as the number of decimal places to be retained and the places for checks.

Expected probits for the experimental points are read from the provisional line, to the nearest 0.1 probit, weighting coefficients are found from a table such as Table 5.1, and weights, nw, thus arrived at. We write x = log-dose. Summation over the different doses gives Snw, $Snwx$ and $Snwx^2$. Probits of a special kind known as working probits, y, are found with the help of tables (Finney, 1971) not given here; each working probit is derived from the expected probit and the observed per cent response, and is usually near in value to the empirical probit. Sums $Snwy$, $Snwxy$ and $Snwy^2$ are computed, together with

$$\bar{x} = Snwx/Snw, \tag{5.4}$$

$$\bar{y} = Snwy/Snw. \tag{5.5}$$

Table 5.1 Weighting coefficients for probit analysis.

	100 C										
Y	0	1	2	3	4	5	6	7	8	9	10
3.0	.131	.091	.069	.056	.046	.040	.034	.030	.027	.025	.022
3.1	.154	.114	.090	.074	.063	.054	.048	.043	.038	.035	.032
3.2	.180	.140	.115	.097	.083	.073	.065	.058	.053	.048	.044
3.3	.208	.169	.142	.123	.107	.095	.085	.077	.070	.065	.059
3.4	.238	.201	.173	.152	.135	.121	.110	.100	.092	.085	.078
3.5	.269	.234	.206	.184	.166	.151	.138	.127	.117	.108	.101
3.6	.302	.268	.241	.218	.199	.183	.169	.156	.145	.136	.127
3.7	.336	.304	.277	.255	.235	.218	.202	.189	.177	.166	.156
3.8	.370	.340	.315	.292	.272	.254	.238	.224	.211	.199	.188
3.9	.405	.377	.352	.330	.310	.292	.275	.260	.247	.234	.222
4.0	.439	.412	.389	.367	.347	.329	.313	.297	.283	.270	.258
4.1	.471	.447	.424	.404	.384	.367	.350	.335	.320	.307	.294
4.2	.503	.480	.458	.439	.420	.403	.386	.371	.356	.343	.330
4.3	.532	.510	.490	.471	.453	.437	.421	.405	.391	.377	.364
4.4	.558	.538	.519	.501	.484	.468	.453	.438	.424	.410	.397
4.5	.581	.563	.545	.528	.512	.496	.481	.467	.453	.440	.427
4.6	.601	.583	.567	.551	.536	.521	.507	.493	.480	.467	.454
4.7	.616	.600	.585	.570	.556	.542	.528	.515	.502	.489	.477
4.8	.627	.613	.598	.584	.571	.558	.545	.532	.520	.508	.496
4.9	.634	.621	.607	.594	.582	.569	.557	.545	.533	.522	.511
5.0	.637	.624	.612	.600	.588	.576	.565	.553	.542	.531	.521
5.1	.634	.623	.611	.600	.589	.578	.567	.557	.546	.536	.526
5.2	.627	.617	.606	.596	.585	.575	.565	.555	.546	.536	.526
5.3	.616	.606	.596	.587	.577	.568	.558	.549	.540	.531	.522
5.4	.601	.591	.582	.573	.565	.556	.547	.539	.530	.522	.513
5.5	.581	.573	.564	.556	.548	.540	.532	.524	.516	.508	.501
5.6	.558	.550	.543	.535	.528	.520	.513	.505	.498	.491	.484
5.7	.532	.525	.518	.511	.504	.497	.490	.484	.477	.470	.464
5.8	.503	.496	.490	.484	.477	.471	.465	.459	.453	.447	.440
5.9	.471	.466	.460	.454	.449	.443	.437	.432	.426	.420	.415
6.0	.439	.433	.428	.423	.418	.413	.408	.403	.398	.392	.387
6.1	.405	.400	.395	.391	.386	.381	.377	.372	.368	.363	.359
6.2	.370	.366	.362	.358	.354	.350	.345	.341	.337	.333	.329
6.3	.336	.332	.328	.325	.321	.317	.314	.310	.306	.303	.299
6.4	.302	.299	.295	.292	.289	.286	.282	.279	.276	.273	.269
6.5	.269	.266	.263	.260	.258	.255	.252	.249	.246	.243	.240
6.6	.238	.235	.233	.230	.228	.225	.223	.220	.218	.215	.213
6.7	.208	.206	.203	.201	.199	.197	.195	.193	.190	.188	.186
6.8	.180	.178	.176	.174	.172	.171	.169	.167	.165	.163	.161
6.9	.154	.153	.151	.150	.148	.146	.145	.143	.142	.140	.139
7.0	.131	.130	.128	.127	.126	.124	.123	.122	.120	.119	.118

After finding

$$S_{xx} = Snw(x-\bar{x})^2 = Snwx^2 - (Snwx)^2/Snw, \quad (5.6)$$

$$S_{xy} = Snwxy - (Snwx)(Snwy)/Snw, \quad (5.7)$$

then follow the relations

$$b = S_{xy}/S_{xx} \quad (5.8)$$

and

$$Y = \bar{y} + b(x-\bar{x}), \quad (5.9)$$

which is the desired regression line. The steps leading to it show the analogies of the computations with simple regression analysis. The log-ED50, m, is estimated as

$$m = \bar{x} + (5-\bar{y})/b.$$

Provided that the data are not heterogeneous (see § 5.4), (*5.10*) and (*5.11*) estimate variances.

$$V(m) = b^{-2}[1/Snw + (m-x)^2/Snw(x-\bar{x})^2] \quad (5.10)$$

$$V(b) = 1/S_{xx}. \quad (5.11)$$

In fact the second term within the square brackets of (*5.10*) can often be neglected.

A second cycle of computations would begin from expected probits based on the regression equation calculated in the first, and would proceed in the same way as the first, though with additional decimal places as appropriate.

5.3 Shortened approximate analysis

In eye-fitting of a probit line, the differences between weights of points can be allowed for to some extent, and the fitting will often give useful estimates of the parameters of the line. Such fitting can be followed by brief computations, namely the first part of a first cycle, which yield approximations to the variances of the log-ED50 and slope. A numerical example now shows the procedure.

Table 5.2 gives a set of data relating mortality of granary weevils to dose of an insecticide, malathion. The basic data are doses, z, numbers of subjects, n, and numbers dead, r.

Fig. 5.1 shows empirical probits from the table plotted against x, with an eye-fitted line. The Y in the table have been read from the line. The weighting coefficients, w, are from Table 5.1 and the remaining quantities follow by calculation.

Table 5.2 Toxicity to granary weevils, *Sitophilus granarius*, of malathion (control mortality zero). Computations for shortened analysis.

z	x*	n	r	Mort. %	Emp. Probit	Y	w	nw	nwx
.16	.20	120	3	2.5	3.04	3.1	.154	18.5	3.700
.22	.34	120	11	9	3.66	4.0	.439	52.7	17.918
.31	.49	119	53	45	4.87	4.8	.627	74.6	36.554
.43	.63	120	91	76	5.71	5.6	.558	67.0	42.210
.60	.78	119	107	90	6.28	6.3	.336	40.0	31.200
$Snwx^2 = 75.672$								252.8	131.582

* $x = 1 + \log z$

From the line, the log-LD50 and slope are $m \simeq 0.53$ and $b \simeq 5.52$. Since m is near $\bar{x}$ we can use the simplified expression

$$V(m) \simeq 1/b^2 Snw,$$

which in this example gives

$$\begin{aligned} V(m) &\simeq 1/5.52^2 \times 252.8 \\ &= 1.30 \times 10^{-4}. \end{aligned}$$

Obviously this estimate depends upon a reasonable estimate for b. From (*5.6*) we have

$$\begin{aligned} S_{xx} &= 75.762 - 131.582^2/252.8 \\ &= 7.274, \end{aligned}$$

so that from (*5.11*)

$$V(b) \simeq 0.137.$$

If a slightly less accurate estimate is acceptable, calculation of $V(b)$ can under suitable conditions be shortened. The conditions, often met, are that there are at least five treated groups of subjects, that the groups contain equal numbers of subjects or nearly so, and that the log-doses are equally spaced. If so, the nwx and nwx^2 need not be computed, and we can use the relation.

$$V(b) \simeq 12(k-1)/FL^2 Snw(k+1). \qquad (5.12)$$

Here k is the number of dose levels, F is a factor determined from Table 5.3, and L is the difference between the highest and lowest log-doses. The value of F depends upon the range of expected probits, Y, upon the control response, C, and upon the number of dose levels. Interpolation in Table 5.3 will in general be required.

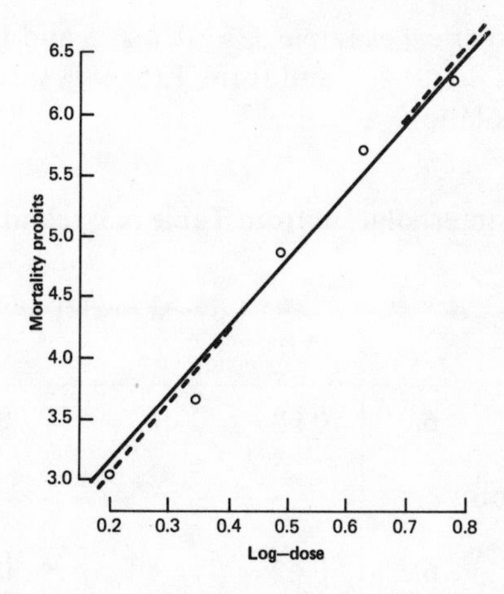

Fig. 5.1 Probit mortality plotted against log-dose for the data of Table 5.2. Fitting by eye (continuous line) and computer (broken line).

Table 5.3 Factors, F, for use with (*5.12*) in approximating $V(b)$.

Probit range	$C = 0$ 5 levels	$C = 0$ 9 levels	$C = 0.10$ 5 levels	$C = 0.10$ 9 levels
3.0 to 5.0	.75	.78	.52	.57
3.0 to 5.5	.71	.75	.52	.56
3.0 to 6.0	.68	.71	.51	.55
3.0 to 6.5	.62	.66	.49	.53
3.0 to 7.0	.54	.59	.43	.48
3.5 to 5.5	.84	.86	.71	.74
3.5 to 6.0	.79	.81	.68	.71
3.5 to 6.5	.72	.75	.62	.67
3.5 to 7.0	.62	.66	.54	.60
4.0 to 6.0	.87	.89	.83	.81
4.0 to 6.5	.79	.81	.75	.78
4.0 to 7.0	.68	.71	.65	.69
4.5 to 6.5	.84	.86	.83	.85
4.5 to 7.0	.71	.75	.71	.74
5.0 to 7.0	.75	.78	.75	.78

For the present numerical example, C = 0, $k = 5$ and L = 0.58. The range of expected probits is 3.1 to 6.3, and from Table 5.3 we construct Table 5.4 preparatory to interpolation.

Table 5.4 Values for interpolation from Table 5.3, relating to the example of Table 5.2.

		Lowest probit			
		3.0		3.5	
Highest probit	6.0	0.68		0.79	
					0.6
		– – – – –	– – – – – – –	– – – –	6.3
					0.4
	6.5	0.62		0.72	
			0.2 ¦ 0.8		
			3.1		

Horizontal interpolation by inspection gives:

		Lowest probit	
		3.1	
Highest probit	6.0	0.70	
			0.6
		– – – – – –	6.3
			0.4
	6.5	0.64	

Vertical interpolation by inspection now gives the required value of F, namely 0.66. However, interpolations by inspection may sometimes give values for F wrong by 0.01 or 0.02, and ordinary double-linear interpolation is safer. Referring back to Table 5.4, for $Y = 3.1$ to 6.3 we compute

$$\begin{aligned} F &= 0.68 \times 0.4 \times 0.8 + 0.79 \times 0.2 \\ &\quad + 0.62 \times 0.6 \times 0.8 + 0.72 \times 0.6 \times 0.2 \\ &= 0.66. \end{aligned}$$

We now have

$$\begin{aligned} V(b) &\simeq 12 \times 4/0.66 \times 0.58^2 \times 252.8 \times 6 \\ &= 0.143, \end{aligned}$$

to be compared with value of 0.137 above.

The estimates of m and b, with their standard errors, from the shortened analysis can now be compared with estimates by computer (Table 5.5). Possibly the agreement between the results from the two methods of calculation is here better than it often might be in analyses of comparable

Table 5.5 Estimates of log-LD50 and slope by short computations and electronic computer. Data of Table 5.2.

	Short method	Computer
Log-LD50	0.53 ± 0.01	0.533 ± 0.011
Slope	5.52 ± 0.37	5.983 ± 0.399
	(Approx. ± 0.38)	

data, and certainly better than for data in which the subjects were substantially fewer. However, the main object of the present section is to describe the shortened, approximate method of analysis.

5.4 Goodness of fit

After a probit line has been fitted, the departures of the observed responses from those calculated from the line can be assessed by a Pearson chi-square test. For this we evaluate

$$X^2 = \mathrm{S}[(r - nP)^2/nPQ] \qquad (5.13)$$

Summation is over the treated groups. In each group, r is the observed number of subjects responding (without correction for control response); n is as before the number of subjects in the group; P is the proportion expected to respond; and $Q = 1 - P$. To find P, the probit response is calculated from the probit-log-dose regression line, and is transformed to a proportion by tables; this proportion is anti-corrected for control response (for example, referring to Table 1.1, a proportion of 0.368 from the probit with $C = 0.05$ would give $P = 0.400$). In essence, the test compares number of subjects observed to respond, r, with numbers expected to, nP.

The value of X^2 is referred to a table of chi-square, with degrees of freedom equal to the number of dose levels, k, less two (since two parameters have been estimated). If the probability arrived at is greater than 0.05, the fit of the observations to the probit line is regarded as adequate, and (*5.10*) and (*5.11*) can be used directly for estimation of variances. If the probability is less than 0.05 the data are said to be heterogeneous, and sound estimates of variances may be difficult to obtain (see Finney, 1971). If the high value of X^2 is merely due to scatter of points, variances can be estimated by multiplying values from (*5.10*) and (*5.11*) by a heterogeneity factor, $X^2/(\mathrm{k}-2)$. If, however, probit response is really curvilinear in log-dose, use of the factor is not justifiable.

As a numerical example Table 5.6, shows the calculation of X^2 for goodness of fit relating to the data of Table 5.2. Expected probits, Y, were found from the line.

$$Y = 1.811 + 5.983\,x \qquad (5.14)$$

Table 5.6 Calculations of X^2 for the data of Table 5.2, testing the fit of the computer-fitted line (Table 5.5).

n	x	Y	P	nP	r	nPQ	X^2 contrib.
120	.204	3.03	.025	3.0	3	2.92	0.00
120	.342	3.86	.128	15.4	11	13.39	1.45
119	.491	4.75	.400	47.6	53	28.56	1.02
120	.633	5.60	.727	87.2	91	23.82	0.61
119	.778	6.47	.929	110.6	107	7.85	1.65
							$X^2 = 4.73$

fitted by computer (see Table 5.5). Referred to a table of chi-square, the X^2 of 4.73 with 3 degrees of freedom gives a probability of 0.20, so that (*5.10*) and (*5.11*) give variances, without a heterogeneity factor. The eye-fitted line (Table 5.5) for the same data yields $X^2 = 5.61$ and a probability of 0.13.

Where some expectations (i.e. nP or nQ) are less than five, a recommendation that often used to be made was that some groups should be combined before calculating X^2. However, Gough (personal communication) has investigated the question of low expectations in probit and logit analysis. His findings indicate that the Pearson test is normally satisfactory even if some expectations are as low as one, and possibly lower. He gives a criterion for deciding whether the Pearson test is applicable, and, where it is not, a modification suitable if expectations exceed 1/4.

6 Calculations in logit analysis

6.1 Introduction

The calculations of logit analysis are in most respects similar to those of probit analysis. We begin by plotting the empirical logits, l, of the observed responses (see Table 4.1) against the corresponding log-doses, x. If the points fit a straight line reasonably well (see § 5.1), we draw a line by eye. The corresponding formal relation is

$$L = \theta + \phi x, \qquad (6.1)$$

where L is the logit of the response at log-dose x. This is also equation (*4.2*). The slope, f, of the line drawn is an estimate of ϕ, which, taken with coordinates of any point on the line, leads to an estimate, t, of θ; $-\theta/\phi$ is here the log-ED50 (contrast probit analysis – (*5.2*)). As in probit analysis, fitting a line by eye may be adequate for some purposes, but an objective method eliminates the subjective element in fitting by eye and provides standard errors of estimates. We now discuss two such methods, namely minimum logit chi-square and maximum likelihood.

6.2 Minimum logit chi-square

Berkson proposed this procedure and has described it in many publications (e.g. 1949, 1953). It is usually a good procedure, in that it normally utilizes nearly all the information in the data, and, usefully, it requires no iteration. However, groups in which the response is zero or 100% cannot be included fully satisfactorily, and so, if an appreciable amount of the data resides in groups with these extreme responses, the method is contra-indicated. The object is to minimize deviations of the empirical logits from the expected logits given by (*6.1*). Each weight is the product of a weighting coefficient, w, and the number of subjects in the group, n. When there is zero response among the controls, the weighting coefficient w is expressed in terms of the empirical response rate, p, by the simple formula

$$w = p(1-p). \qquad (6.2)$$

For a control response rate C, the general expression is

$$w = p^2(1-p)/\{p + C/(1-C)\}. \qquad (6.3)$$

Table 6.1 Weighting coefficients for logit analysis.

	100 C										
p	0	1	2	3	4	5	6	7	8	9	10
.02	.020	.013	.010	.008	.006	.005	.005	.004	.004	.003	.003
.04	.038	.031	.025	.022	.019	.017	.015	.013	.012	.011	.010
.06	.056	.048	.042	.037	.033	.030	.027	.025	.023	.021	.020
.08	.074	.065	.059	.053	.048	.044	.041	.038	.035	.033	.031
.10	.090	.082	.075	.069	.064	.059	.055	.051	.048	.045	.043
.12	.106	.097	.090	.084	.078	.073	.069	.065	.061	.058	.055
.14	.120	.112	.105	.099	.093	.088	.083	.078	.074	.071	.067
.16	.134	.126	.119	.113	.107	.101	.096	.091	.087	.083	.079
.18	.148	.140	.133	.126	.120	.114	.109	.104	.100	.095	.091
.20	.160	.152	.145	.139	.132	.127	.121	.116	.112	.107	.103
.22	.172	.164	.157	.150	.144	.138	.133	.128	.123	.118	.114
.24	.182	.175	.168	.162	.155	.150	.144	.139	.134	.129	.125
.26	.192	.185	.178	.172	.166	.160	.154	.149	.144	.139	.135
.28	.202	.195	.188	.182	.175	.170	.164	.159	.154	.149	.144
.30	.210	.203	.197	.190	.184	.179	.173	.168	.163	.158	.153
.32	.218	.211	.205	.198	.193	.187	.181	.176	.171	.166	.162
.34	.224	.218	.212	.206	.200	.194	.189	.184	.179	.174	.169
.36	.230	.224	.218	.212	.206	.201	.196	.191	.186	.181	.176
.38	.236	.229	.224	.218	.212	.207	.202	.197	.192	.187	.182
.40	.240	.234	.228	.223	.217	.212	.207	.202	.197	.192	.188
.42	.244	.238	.232	.227	.222	.216	.211	.207	.202	.197	.193
.44	.246	.241	.235	.230	.225	.220	.215	.210	.206	.201	.197
.46	.248	.243	.238	.233	.228	.223	.218	.213	.209	.204	.200
.48	.250	.244	.239	.234	.230	.225	.220	.216	.211	.207	.203
.50	.250	.245	.240	.235	.231	.226	.222	.217	.213	.209	.205
.52	.250	.245	.240	.236	.231	.227	.222	.218	.214	.210	.206
.54	.248	.244	.239	.235	.231	.226	.222	.218	.214	.210	.206
.56	.246	.242	.238	.234	.229	.225	.221	.217	.213	.209	.206
.58	.244	.239	.235	.231	.227	.223	.219	.216	.212	.208	.204
.60	.240	.236	.232	.228	.224	.221	.217	.213	.210	.206	.202
.62	.236	.232	.228	.224	.221	.217	.214	.210	.207	.203	.200
.64	.230	.227	.223	.220	.216	.213	.210	.206	.203	.200	.196
.66	.224	.221	.218	.214	.211	.208	.205	.201	.198	.195	.192
.68	.218	.214	.211	.208	.205	.202	.199	.196	.193	.190	.187
.70	.210	.207	.204	.201	.198	.195	.192	.190	.187	.184	.181

p	100 C 0	1	2	3	4	5	6	7	8	9	10
.72	.202	.199	.196	.193	.191	.188	.185	.183	.180	.177	.175
.74	.192	.190	.187	.185	.182	.180	.177	.175	.172	.170	.167
.76	.182	.180	.178	.175	.173	.171	.168	.166	.164	.161	.159
.78	.172	.169	.167	.165	.163	.161	.159	.156	.154	.152	.150
.80	.160	.158	.156	.154	.152	.150	.148	.146	.144	.142	.140
.82	.148	.146	.144	.142	.140	.139	.137	.135	.133	.132	.130
.84	.134	.133	.131	.130	.128	.126	.125	.123	.122	.120	.119
.86	.120	.119	.118	.116	.115	.113	.112	.111	.109	.108	.107
.88	.106	.104	.103	.102	.101	.100	.098	.097	.096	.095	.094
.90	.090	.089	.088	.087	.086	.085	.084	.083	.082	.081	.080
.92	.074	.073	.072	.071	.070	.070	.069	.068	.067	.066	.066
.94	.056	.056	.055	.055	.054	.053	.053	.052	.052	.051	.050
.96	.038	.038	.038	.037	.037	.036	.036	.036	.035	.035	.034
.98	.020	.019	.019	.019	.019	.019	.018	.018	.018	.018	.018

Table 6.1 is entered with the empirical response rate, p, and control mortality rate, C, whence the weighting coefficient required can be obtained by interpolation as described in Chapter 5. Alternatively, the coefficient may be calculated directly from the formula.

The main part of the analysis consists of a weighted regression procedure without iteration. Provided that there are no groups with a response of zero or 100%, all the expressions from (*5.4*) to (*5.9*) inclusive remain valid with y replaced by an empirical logit l and w defined by (*6.2*) or (*6.3*) above. The fitted line is

$$l = t + fx, \tag{6.4}$$

where

$$f = S_{xl}/S_{xx} \tag{6.5}$$

and

$$t = \bar{l} - f\bar{x}. \tag{6.6}$$

A change in the expression for the log-ED50 arises when logits are used instead of probits (see § 6.1), because the former are centred on zero and the latter on the value 5. The log-ED50 is now estimated by

$$m = \bar{x} - \bar{l}/f. \tag{6.7}$$

When the experiment contains groups where the response is zero or 100%, we recommend that such groups are omitted when calculating the line or included

Table 6.2 Toxicity of malathion to granary weevils. Principal calculations for minimum logit chi-square analysis.

x	n	p	l	w	nw	nwx
.20	120	.025	−3.66	.024	2.9	0.580
.34	120	.092	−2.29	.084	10.1	3.434
.49	119	.445	−0.22	.247	29.4	14.406
.63	120	.758	+1.14	.183	22.0	13.860
.78	119	.899	+2.19	.091	10.8	8.424

$Snw = 75.2$ $\quad Snwx = 40.704$ $\quad Snwl = 8.521$
$Snwx^2 = 23.645$ $\quad Snwxl = 21.093$ $\quad Snwl^2 = 173.625$
$\bar{x} = 0.541$ $\quad \bar{l} = 0.113$
$S_{xx} = 1.613$ $\quad S_{xl} = 16.481$ $\quad S_{ll} = 172.659$

as proportions of $1/2n$ or $(1-1/2n)$ (Berkson, 1953). They must, however, be included when testing for goodness of fit as described in § 6.4.

Table 6.2 shows the principal calculations required when the minimum logit chi-square procedure is applied to the data on toxicity of malathion to granary weevils. Here x and n are the same as in Table 5.2, p is the empirical mortality expressed as a proportion, l is the empirical logit, and w is the weighting coefficient given by (*6.2*), the control response being zero. The estimated values of ϕ and θ are respectively

$$f = 16.481/1.613 = 10.2$$

and

$$t = 0.113 - 10.2(0.541) = -5.41.$$

Thus the log ED50 is estimated by

$$m = 0.541 - 0.113/10.2 = 5.41/10.2 = 0.530$$

The standard errors of m *and* f are calculated from (*5.10*) and (*5.11*), whence

$$\text{s.e.}(m) = 0.011, \ \text{s.e.}(f) = 0.79.$$

6.3 Maximum likelihood

This procedure is another objective method of estimating a linear relationship between logit and log-dose, as well as finding standard errors of the estimates. The complete process is one of successive approximations, and parallels what has already been described for probits in Sections 5.1 and 5.2. After an eye-fitted line is drawn through the experimental points, expected logits are read from the line and converted into working logits by special tables (Finney,

1964). A weighted regression analysis then gives an approximation to the maximum likelihood estimates required and completes the first cycle of the calculations. The second cycle begins with the expected logits obtained at the end of the first cycle, and repeats the regression calculations with modified numbers. This iterative procedure is conveniently implemented on an electronic computer (but see § 6.5).

A shortened version of the full procedure is obtained by using expected logits instead of working logits, and carrying out just one cycle of the calculations. This represents a modified version of minimum logit chi-square, in which empirical logits are replaced by expected logits and the weighting coefficients are changed accordingly.

Table 6.3 Toxicity of malathion to granary weevils. Principal calculations for shortened maximum likelihood logit analysis.

x	n	p	l	L	w	nw	nwx
.20	120	.025	−3.66	−3.40	.031	3.7	0.740
.34	120	.092	−2.29	−1.83	.119	14.3	4.862
.49	119	.445	−0.22	−0.37	.242	28.8	14.112
.63	120	.758	1.14	0.81	.213	25.6	16.128
.78	119	.899	2.19	2.31	.082	9.8	7.644

$Snw = 82.2$	$Snwx = 43.486$	$SnwL = -6.031$
$Snwx^2 = 24.839$	$SnwxL = 14.086$	$SnwL^2 = 163.694$
$\bar{x} = 0.529$		$\bar{l} = -0.073$
$S_{xx} = 1.834$	$S_{xl} = 17.277$	$S_{ll} = 163.252$

Table 6.3 shows the principal calculations required when the shortened maximum likelihood logit analysis is applied to the data on toxicity of malathion to granary weevils. Here x, n, p and l are the same as in Table 6.2, L is the expected logit read from the eye-fitted line, and w is the weighting coefficient corresponding to L. The estimates with their standard errors in brackets are now

$$m = 0.537 \pm 0.012$$

and

$$f = 9.42 \pm 0.74.$$

6.4 Goodness of fit

After fitting a linear relationship between logit and log-dose, the observed frequencies are compared with the expected frequencies obtained from the line. There are two methods of comparison: individual and overall, each of which can be implemented by straightforward methods based on the analysis of the previous sections.

Individual comparisons are made by calculation of standardized residuals, which are the deviations between observed and expected frequencies measured on a scale which allows for variations in the mortality rate. A typical standardized residual is defined by

$$d = (r - nP)/(nPQ)^{\frac{1}{2}}. \tag{6.8}$$

The purpose of calculating these quantities is to see where the differences lie between the experimental data and the assumption of linearity. Inspection of the pattern of residuals will suggest further analyses when there are departures from linearity.

An overall comparison between observed and expected frequencies is made by calculating Pearson's statistic

$$X^2 = \mathrm{S}(r - nP)^2/nPQ. \tag{6.9}$$

This is also the sum of squares of the standardized residuals, so that

$$X^2 = \mathrm{S}d^2. \tag{6.10}$$

The significance of a value of X^2 is assessed by reference to a table of chi-square with $(k-2)$ degrees of freedom, where k is the number of dose levels.

Further comments on how to compute X^2, interpret significance levels, allow for heterogeneity, and deal with small frequencies, are made in Section 5.4.

The methods are illustrated by further analysis of the data on toxicity of malathion to granary weevils. Logit analysis by the shortened maximum likelihood procedure leads to the line

$$L = 9.42x - 5.056.$$

Table 6.4 presents an analysis of the standardized residuals. None of them exceeds two in absolute value and is therefore large enough to merit comment. The value of Pearson's statistic is

$$X^2 = 5.79.$$

A probability of 0.12 is obtained from the table of chi-square with 3 degrees of freedom, and confirms that the fit is satisfactory.

6.5 Use of portable programmable calculators

Whereas probit analysis is difficult for them, modern portable programmable electric calculators are most effective for logit analysis, as they compute logits and their weights with ease. A comparatively modest type with 100 program

Table 6.4 Toxicity of malathion to granary weevils. Analysis of residuals for shortened maximum likelihood logit analysis.

x	n	r	L	P	nP	$(nPQ)^{\frac{1}{2}}$	d
.20	120	3	−3.172	.040	4.8	2.15	−0.84
.34	120	11	−1.853	.136	16.3	3.75	−1.41
.49	119	53	−0.440	.392	46.6	5.32	1.20
.63	120	91	0.879	.707	84.8	4.98	1.24
.78	119	107	2.292	.908	108.1	3.15	0.35

steps and ten addressable memory registers is adequate, enabling logit lines and the relevant errors to be calculated rapidly, without reference to tables. Programmed to carry out full maximum likelihood iterations (including working logits and allowance for control response if present), one of these calculators was used to fit lines to the data of Table 6.2. The program was in effect a dual purpose one. Initially, an input of a slightly modified form yielded the line fitted by minimum chi-square, which gave expected logits for a first maximum likelihood iteration. A second, similar iteration resulted in estimates for the parameters unchanged by further iterations. Table 6.5 presents the results, with the previous ones for comparison.

Table 6.5 Logit analysis of the data of Table 6.2.

	Minimum chi-square	Maximum likelihood 1 iteration*	Maximum likelihood 2 iterations†
Log-LD50 ± S.E.	0.530 ± 0.011	0.537 ± 0.012	0.530 ± 0.011
Slope ± S.E.	10.22 ± 0.79	9.42 ± 0.74	10.39 ± 0.79

* Shortened analysis, Table 6.3.
† Following minimum chi-square fitting.

It will be seen that, following the two iterations, the maximum likelihood estimate of the log-LD50 and its standard error were the same as from the computer probit analysis of the same data (Table 5.5), also by maximum likelihood. At the end of Chapter 4 we point out that from theory the slope of a logit regression line can be expected to be about 1.814 times that of a probit regression line for the same data. In Table 5.5 the slope of the computer probit line is 5.98, and $5.98 \times 1.814 = 10.85$; this agrees reasonably well with the comparable slope of 10.39 in Table 6.5.

7 Quantal responses to mixtures of drugs

7.1 Generalities

The greatest caution needs to be exercised in administering to a human being more than one drug at a time. Two drugs may interact and, unless their joint effects are known well, unexpected damage to the patient may result. Since a number of commercial drug preparations include more than one drug, problems may be encountered not infrequently (Burns *et al.*, 1965; Goldstein *et al.*, 1969).

Manufacturers of pesticides, drugs that destroy life selectively, are not subject to so severe a restriction as manufacturers of medicinal preparations. The former can mix active ingredients with some freedom in their formulations, but manufacturing complications generally make them reluctant to do so. Their reluctance appears to be manifested in a simple and interesting statistical effect. Hewlett (1968) examined the numbers of insecticidal formulations, marketed in the U.S.A. in 1957 and 1966, containing 1, 2, 3, 4 . . . active ingredients. In 1957, 1100 formulations were marketed containing one active ingredient, nearly $1100 \times 0.235 = 258$ containing two active ingredients, nearly $1100 \times 0.235^2 = 61$ containing 3 active ingredients, and so on in a descending geometric progression. In 1966, the same type of relation applied, but the factor, 0.262, was significantly higher, so that in the later year the manufacturers seem to have been a little less disinclined to use multiple formulations.

Synergists can be defined as substances pharmacologically inactive when applied alone, but which increase the activities of drugs. Synergists are important in laboratory investigations of drug action (Goldstein *et al.*, 1969; Dyte and Rowlands, 1968, 1970; Albert, 1973), and in practice are used widely in pesticide formulations, especially in association with the insecticidal compounds known as pyrethroids. However, numerous insecticidal formulations each contain two or more substances that are separately insecticidal.

Three topics will be discussed in this chapter: mathematical models for quantal responses to mixtures of drugs, measurement of their joint action, and design of related experiments. Phenomena associated with drug-induced induction of enzymes are not considered.

7.2 Types of model

We shall discuss models for the joint action of two drugs only. Models for three or more drugs need introduce no differences in principle. Any consistent model relating response to the doses z_1 and z_2 of two drugs, A and B, must reduce to the proper relation between response and z_1 as z_2 is made zero, and to that between response and z_2 as z_1 is made zero. The following types of mathematical model can be distinguished.

(1) *Stochastic models.* Puri and Senturia (1972) developed a stochastic model giving a fairly manageable relation between response and dose for one drug. In approaching corresponding models for more than one drug, they ran into mathematical difficulties. Thus stochastic models for joint action have not yet been developed, but in order to see whether such models would be useful, the reasonable course would be to develop some and see.

(2) *Deterministic models.* The numbers of drug molecules interacting with living tissues in pharmacological and toxicological experiments are so vast that deterministic models of drug action can be successful – as in fact they have been. Deterministic models are familiar to pharmacologists and toxicologists, and five types are distinguishable.

(a) Models (mostly semi-qualitative) conceived from isobolograms (§ 7.3).

(b) Quantitative models satisfying minimal general requirements, e.g. giving suitable limiting cases (§ 7.4 to 7.6).

(c) Models based on general biological concepts (§ 7.7 to 7.12).

(d) Models arising from drug-receptor theory (§ 7.13).

(e) Physico-chemically realistic models, based on reactions of drug molecules with tissue components of which the molecular configurations are known. No such model has yet been constructed, although one for the action of acetylcholine probably will be before long.

Hewlett and Plackett (1961) proposed a general method for developing models of joint drug action.

An isobologram for two drugs consists of two axes at right angles along which the doses of the respective drugs are measured, and an isobole, a curve any point on which represents the dose-pair for a chosen level of response (see Figs 7.1 and 7.2). The response could be graded or quantal, but here we are concerned only with the latter. Models of types (b), (c) and (d) give rise to equations relating proportional response, P, to doses z_1 and z_2, of the respective drugs. For given values of the parameters of such an equation, we can construct a set of isoboles for different respective values of P. We can therefore use isobolograms for the graphical representation of models of types (b) and (c), although isobolograms were not originally conceived for this purpose.

The account that follows directs most attention to models of type (c).

7.3 Models from isobolograms

Isoboles have been used in pharmacology for about one hundred years, and Loewe was one of their chief proponents (Loewe and Muishnek, 1926; Loewe, 1953, 1959). The terminology for the models concerned has not been fully standardised, and here we follow that given by Ariëns *et al*, (1976).

Fig. 7.1 illustrates isoboles for the limiting case where one drug, A, is active when alone and the other, B, is not, i.e. it is inert. A dose of A, $z_1 = Z_1$ produces the given effect. If there is *additive action* (sometimes called addition), the isobole (I) is a horizontal line through C, for which $z_1 = Z_1$. Thus here the dose of B does not influence the dose of A required to produce the effect. If B reduces the dose of A required to produce the given effect, the isobole proceeds downward from the point C; synergism (II) occurs. If the

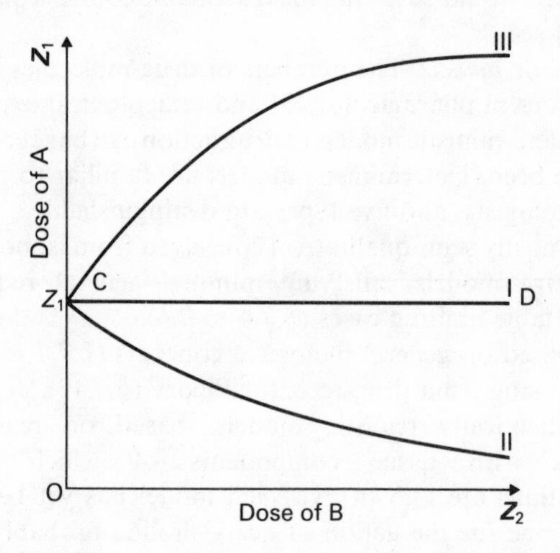

Fig. 7.1 Isoboles where one drug, B, is inert. The line CD(I) is for addition. II is an isobole for synergism; III, one for antagonism.

action of B is opposite, so that the isobole proceeds upwards from C, antagonism (III) occurs.

Fig. 7.2 illustrates isoboles where the two drugs, A and B, are both separately active. The principle is the same as in Fig. 7.1. The isobole (IV) is the straight line CD for *additive action*. Isoboles for potentiation (V) lie below CD, so that here lower doses are necessary for the given effect that if the action is additive. Isoboles (VI) for subadditive action proceed from C and D, but lie within area CDE; A and B each increase the action of the other, but the nett effect of the combination is less than in additive action. If an isobole (VII) proceeding from C to D lies above and to the right of CDE, antagonism occurs. Finally, *coalitive action* occurs if neither drug is separately effective, yet the two together are so; (VIII) is an example of an isobole for this type of

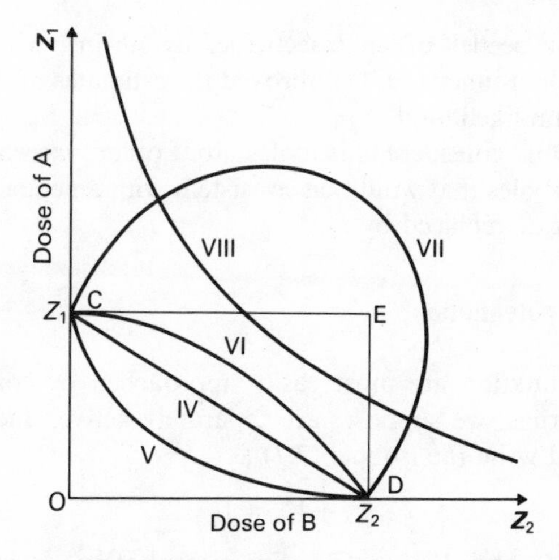

Fig. 7.2 Isoboles where both drugs are separately active. CD(IV) is the isobole for addition; V is one isobole for potentiation; VI, one for subadditive action; VII, one for antagonism, and VIII, one for coalitive action.

action, the isobole never reaching either axis, though otherwise unrestricted in position.

In pharmacology, isobolograms have been invaluable in enabling experimental data to be examined in an organized manner. However, all but one of the models concerned are defined by regions in which isoboles fall, and are therefore not fully defined quantitatively. The model for additive action is fully defined quantitatively, and is given by the equation

$$z_1/Z_1 + z_2/Z_2 = 1. \qquad (7.1)$$

7.4 A model for synergism

A model for synergism proposed by Hewlett (1969a) relates probit response to the dose of a drug, A, with the dose of a synergist, B, inactive when alone. The model is

$$\mathrm{Y} = \alpha + \beta_1 \log z_1 + \beta_2 z_2/(\gamma + z_2) \qquad (7.2)$$

For the drug alone ($z_2 = 0$) probit response is linear in the log-dose of A, and for the synergist alone ($z_1 = 0$) the proportional response is zero. Furthermore, the last term of (*7.2*) is such that, for a fixed dose of A, response increases towards a fixed level as the synergist dose increases indefinitely, which conforms with an expectation that the synergist dose will not increase response indefinitely. Though without a definite biological basis, (*7.2*) gives reasonable limiting cases.

Hewlett (1969a) showed that (*7.2*) adequately represented data for the toxicities to flour beetles of an insecticide, pyrethrins, with a synergist, piperonyl butoxide. Finney (1971) improved the estimates of the parameters by using maximum likelihood.

Hewlett (1969b) considered isoboles for other insecticide-synergist combinations, isoboles that would be consistent with an equation similar to (*7.2*) but with log z_1 replaced by z_1.

7.5 Models for potentiation

Models for potentiation are most easily approached by consideration of isoboles. Both drugs, we suppose, are separately active, and if they were additive we could write the isobole (*7.1*) as

$$V_1 + V_2 = 1 \tag{7.3}$$

where, $V_1 = z_1/Z_1$ and $V_2 = z_2/Z_2$. For potentiation an isobole will be generally like curve (V) in Fig. 7.2, and will be described by a generalization of (*7.3*). Suitable simple generalizations of (*7.3*) may be found by introducing a single parameter, as in

$$V_1^\eta + V_2^\eta = 1 \quad (\eta \leqslant 1) \tag{7.4}$$

$$V_1 + \kappa(V_1 V_2)^{\frac{1}{2}} + V_2 = 1 \quad (\kappa \geqslant 0) \tag{7.5}$$

For discussion of these and comparable relations, see Hewlett (1969b). (*7.4*) fitted data for the joint toxicity to a grass of two strongly potentiating herbicides, and both (*7.4*) and (*7.5*) fitted data for the joint toxicity of two insecticides to flies (Hewlett, 1969b). Both these models are symmetric in V_1 and V_2. A model assymetric in V_1 and V_2 and incorporating two parameters, is

$$V_1 + \mathrm{K} V_1{}^{\mathrm{n}} V_2 + V_2 = 1 \tag{7.6}$$

As symmetric models were inadequate Hewlett (1969c) fitted this model to data for the joint toxicity of two insecticides, thanite and aprocarb. The necessity for an asymmetrical model reflected the probability that thanite intensified biochemically the toxicity of aprocarb, but not *vice versa*.

Equations (*7.4*), (*7.5*) and (*7.6*) represent isoboles, which are, of course, relations between doses of drugs for combinations giving a fixed level of response. In order to obtain the corresponding mathematical models relating probit (or logit) response to doses of the drugs such as to yield the same values of the parameter(s) irrespective of the level of response chosen, all we have to do is to put

$$V_1 = \mathrm{hxp}\ [(Y_1 - Y)/\beta_1] \tag{7.7}$$

$$V_2 = \mathrm{hxp}\ [(Y_2 - Y)/\beta_2] \tag{7.8}$$

Here hxp is read as 'h to the power of' and h is the base of the logarithms (a notation analogous to exp for exponential).

7.6 Coalitive action

The quantitative aspects of coalitive action have been very little investigated. However, Hewlett and Wilkinson (1967) found that graphical fitting of rectangular hyperbolic isoboles for LD50's,

$$z_1 = \alpha + \beta / z_2 \qquad (7.9)$$

was useful in interpreting data for the joint toxicity to flies of carbaryl with various compounds. Carbaryl, though in many circumstances insecticidal on its own, was here by itself insufficiently toxic to give 50% mortality. With very high doses of many of the other compounds the dose of carbaryl was about the same for 50% mortality, which enabled the relative activities of the former to be compared quantitatively by the estimates of $1/\beta$.

The probit plane

$$Y = \alpha + \beta_1 \log z_1 + \beta_2 \log z_2 \qquad (7.10)$$

is another possible model for coalitive action, but this has not been looked into. The isobole here is

$$v {z_1}^{\beta_1} {z_2}^{\beta_2} = 1 \qquad (7.11)$$

7.7 Models from general biological concepts

Taking models proposed by Bliss (1939) as a point of departure, we developed the set of models described below (see Plackett and Hewlett, 1948, 1952, 1967; Hewlett and Plackett, 1959, 1961). Hewlett (1960) and Finney (1971) reviewed respectively the more biological and the more mathematical aspects of these. Plackett and Hewlett (1967) discussed the models developed along more mathematical lines by Ashford and Smith (1965, etc.); who used these models largely for interpretation of data on incidence of pneumoconiosis under different conditions of dust exposure. Sakai (1960) proposed a subtle and elaborate set of models based partly upon Bliss's and our own.

Models of joint action require a general concept of the action of a single drug. A drug is here regarded as selective in its physiological or biochemical action on the individual organism, and is assumed to cause the response by producing changes in a particular physiological system known as the site of action of the drug. Ariëns (1968) has presented a more detailed concept. When two drugs act at a common site, their joint action is said to be *similar*; when at different sites, it is said to be *dissimilar*. Furthermore, drugs A and B are said to *interact* if the presence of A influences the amount of B reaching B's site of action, and/or the effect of this amount. An equivalent statement holds with A and B interchanged. Interaction is a concept that may include synergism, antagonism and coalitive action. Interaction may or may not occur irrespective of whether the joint action is similar or dissimilar, so that four types of joint action can now be distinguished.

	Similar	Dissimilar
Non-interactive	simple similar	independent
Interactive	complex similar	dependent

The above table gives a convenient set of terms, but it should be pointed out that non-interactive action is the limiting case of interactive action, with zero interaction. Moreover, a joint action may be in some sense partially similar, so that similar and dissimilar actions are to be regarded as the extremes of a biological continuum.

In order to construct mathematical models of quantal responses to mixtures, we need to make assumptions concerning the relation between amounts acting, w, and dose, z. We assume that for the individual organism

$$w_1 = \mu_1 z_1^{\eta_1} \qquad w_2 = \mu_2 z_2^{\eta_2} \qquad (\mu, \eta > 0) \tag{7.12}$$

are sufficiently accurately. These relations are flexible. Obviously $w < z$, and Plackett and Hewlett (1952) mention the conditions under which this holds. Although some variation might in fact occur, for simplicity we assume that the values of μ and η do not vary from one individual organism to another.

From (*3.4*), the probit response to a separate drug is

$$Y_i = \alpha_i + \beta_i \log z_i. \quad (\mathrm{i} = 1{,}2) \tag{7.13}$$

Combining this with (*7.2*) we find

$$Y_i = \zeta_i + \lambda_i \log w_i \tag{7.14}$$

where

$$\lambda_i = \beta_i / \eta_i \tag{7.15}$$

As a final preliminary before deriving models of joint action, we introduce the assumption of correlation of the action tolerances for the two drugs. These tolerances, $\check{w}_1$ and $\check{w}_2$, may be correlated to a greater or lesser degree, and even if the joint action is (physiologically) independent their correlation is still a possibility. In allowing for the correlation, the distribution of log $\check{w}_1$ and log $\check{w}_2$ can be assumed to be bivariate normal if probits are used, but a simpler alternative is available if logits are used.

7.8 Simple similar action

When two drugs act non-interactively at the same site, their effects at the site can be expected to be in some sense additive. How, then, can the effect of a drug at the site of action be measured in some way relevant to the quantal response under consideration? In fact a relevant and simple measure is the ratio $w_i/\check{w}_i$. This ratio rises from zero where $w_i = 0$, to unity where $w_i = \check{w}_i$.

Thus writing Pr for 'the probability that', the probability of response to a drug applied singly is

$$P_i = \Pr\{w_i > \check{w}_i\} = \Pr\{w_i/\check{w}_i > 1\} \quad (i = 1,2). \tag{7.16}$$

It is now natural to suppose that the probability of response to w_1 of drug A together with w_2 of drug B is

$$P = \Pr\{w_1/\check{w}_1 + w_2/\check{w}_2 > 1\} \tag{7.17}$$

A resemblance of (*7.17*) to (*7.1*) is noticeable. There is another route to (*7.17*). We may suppose that w_2 of B acting will be equivalent, i.e. have the same effect as $w_2(\check{w}_1/\check{w}_2)$ of A, the term in brackets being, so to speak, a potency of B relative to A at the site of action for the individual organism concerned. From (*7.16*) we have for drug A

$$P_1 = \Pr\{w_1 > \check{w}_1\} \tag{7.18}$$

and hence we obtain

$$P = \Pr\{w_1 + w_2(\check{w}_1/\check{w}_2) > \check{w}_1\} \tag{7.19}$$

which gives (*7.17*) immediately on dividing each side of the unequality by $\check{w}_1$. Equation (*7.17*) also arises from receptor theory (see § 7.13).

7.9 Simple similar action with positive correlation of tolerances

Equation (*7.17*) is a general model for simple similar action, too general to permit direct evaluation of P. Such evaluation in general necessitates an assumption as to the joint distribution of $\check{w}_1$ and $\check{w}_2$, but this will be discussed in Section 7.11. In many biological situations simple similar action can be expected to be associated with complete positive correlation of $\check{w}_1$ with $\check{w}_2$ and of $\check{z}_1$ with $\check{z}_2$, and we now derive some fairly simple models resulting from the assumption of the complete positive correlation.

From (*7.12*) and (*7.17*) we have

$$P = \Pr\left\{\left(\frac{z_1}{\check{z}_1}\right)^{\eta_1} + \left(\frac{z_2}{\check{z}_2}\right)^{\eta_2} > 1\right\} \tag{7.20}$$

Now where $\check{z}_1$, and $\check{z}_2$ are completely and positively correlated, an individual with a tolerance $\check{z}_1$ equal to, say, the ED50 for drug A alone, will have a tolerance $\check{z}_2$ equal to the ED50 for drug B alone. Similar considerations hold for any other particular ED. Thus if Z_1 and Z_2 are the respective particular values of z_1 and z_2 for the separate drugs giving a particular probit, or logit, of response (Y or l), the relation between the doses of z_1 and z_2 of the drugs applied jointly that give the response Y or l is

$$\left(\frac{z_1}{Z_1}\right)^{\eta_1} + \left(\frac{z_2}{Z_2}\right)^{\eta_2} = 1 \tag{7.21}$$

We can now reach the probit forms of (*7.21*). Equation (*7.13*) gives

$$z_i = \text{hxp}\,[(Y_i - \alpha_i)/\beta_i] \quad (i = 1,2) \qquad (7.22)$$

where hxp means 'h to the power of', h being the base of logarithms. Also

$$Z_i = \text{hxp}\,[(Y - \alpha_i)/\beta_i], \quad (i = 1,2) \qquad (7.23)$$

Y being the probit of P. Applying (*7.15*) and (*7.21*)–(*7.23*) to (*7.21*) gives

$$\text{hxp}\,[(Y_1 - Y)/\lambda_1] + \text{hxp}\,[(Y_2 - Y)/\lambda_2] = 1 \qquad (7.24)$$

which is the probit form of (*7.21*).

We may now set out four special cases of (*7.17*) found by selecting particular values for λ_i and η_i.

(*a*) Suppose $\lambda_1 = \lambda_2 = \lambda$, $\eta_1 \neq \eta_2$ so that $\beta_1 \neq \beta_2$. Then

$$\text{hxp}\,(Y/\lambda) = \text{hxp}\,(Y_1/\lambda) + \text{hxp}\,(Y_2/\lambda) \qquad (7.25)$$

which can be expressed in terms of doses as

$$Y = \lambda \log\,(\text{h}^{\alpha_1/\lambda} z_1^{\beta_1/\lambda} + \text{h}^{\alpha_2/\lambda} z_2^{\beta_2/\lambda}) \qquad (7.26)$$

(*b*) Suppose $\lambda_1 = \lambda_2 = \lambda$, $\eta_1 = \eta_2 \neq 1$ so that $\beta_1 = \beta_2 = \beta$. Then

$$\text{hxp}\,[(Y_1 - Y)/\lambda] + \text{hxp}\,[(Y_2 - Y)/\lambda] = 1 \qquad (7.27)$$

Y is obtained explicitly here, as

$$Y = \lambda \log\,[\text{hxp}(Y_1/\lambda) + \text{hxp}(Y_2/\lambda)] \qquad (7.28)$$

If in the mixture doses the drugs are in just one proportion, $\pi_1 : \pi_2$ $(\pi_1 + \pi_2 = 1)$, then

$$Y = \lambda \log\,[\pi_1{}^{\eta}\, \text{hxp}(\alpha_1/\lambda) + \pi_2{}^{\eta}\, \text{hxp}(\alpha_2/\lambda)] + \beta \log z, \qquad (7.29)$$

z being the total dose.

(*c*) Suppose $\lambda_1 = \lambda_2 = \lambda$, $\eta_1 = \eta_2 = 1$ so that $\lambda = \beta_1 = \beta_2 = \beta$.

Here $\text{hxp}\,[(Y_1 - Y)/\beta] + \text{hxp}\,[(Y_2 - Y)/\beta] = 1 \qquad (7.30)$

However, (*7.30*) becomes

$$Y = \alpha_1 + \beta \log(z_1 + k z_2) = \alpha_2 + \beta \log(z_1/k + z_2) \qquad (7.31)$$

where $k = \text{hxp}[(\alpha_2 - \alpha_1)/\beta]$ i.e. the potency of the first drug relative to the second. Equation (*7.31*) was the equation for simple similar action used by Bliss (1939) and more fully discussed by Finney (1971).

(*d*) Finally suppose $\lambda_1 \neq \lambda_2$, $\eta_1 = \eta_2 = 1$ so that $\beta_1 \neq \beta_2$. This gives

$$\text{hxp}[(Y_1 - Y)/\beta_1] + \text{hxp}[(Y_2 - Y)/\beta_2] = 1 \qquad (7.32)$$

Both (*7.26*) and (*7.32*) are of course models for simpler similar action with complete positive correlation of tolerances, such as to permit non-parallel probit-log-dose lines for the separate drugs. In fitting (*7.26*) to data five parameters are to be estimated; in fitting (*7.32*), only four.

The logit forms of (*7.22*)–(*7.32*) are merely these equations with Y_1, Y_2, Y replaced by l_1, l_2, l. However in view of (*4.1*) these logit equations can be expressed in terms of proportions. For example if the logarithms are natural, the logit equation corresponding to (*7.32*) can be expressed as

$$(P_1Q/PQ_1)^{1/\beta_1}+(P_2Q/PQ_2)^{1/\beta_2} = 1 \qquad (7.33)$$

Here $Q_1 = 1-P_1$, $Q_2 = 1-P_2$ and $Q = 1-P$.

7.10 Independent action

If the action of two drugs A and B is independent, an individual organism receiving both will respond if it receives an above-tolerance dose of A, or an above-tolerance dose of B, or both; it will not respond if it receives below-tolerance doses of both. Having derived them in more detail in 1948, we now derive only briefly the models for independent action assuming complete positive, zero, or complete negative correlation of tolerances.

(a) *Complete positive correlation.* Suppose that we take a randomly selected group of organisms and imagine them to be arranged in a line in order of ascending order of tolerance to A from left to right. If the tolerances for the two drugs are completely and positively correlated, the organisms of the group will be similarly arranged in order of ascending tolerance to B. Thus if A and B are administered together (but not all the organisms respond), portions of the group extending from the left end of the line will respond, and we see that

$$P = P_1 \text{ or } P_2 \qquad (7.34)$$

whichever is greater.

(b) *Zero correlation.* If the tolerances to A and B are uncorrelated, arrangement of a randomly selected group of organisms in ascending order of tolerance to A will leave the organisms in random order with respect to tolerance to B. Thus if a proportion P_1 responds to A, of the remaining proportion $(1-P_1)$ a proportion P_2 will respond to B and

$$P = P_1+P_2(1-P_1) = P_1+P_2-P_1P_2 \qquad (7.35)$$

(c) *Negative correlation.* Here the arrangement of the organisms in ascending order of tolerance to A results in a descending order of tolerance to B. The organisms responding to A and B will (assuming not all respond) occur at opposite ends of the line so

$$P = P_1 + P_2 \quad (7.36)$$

$$\text{or } P = 1$$

whichever is less.

7.11 General models for non-interactive action

A general model for non-interactive action must give models for simple similar and independent actions as extreme special cases, and must enable the possibility of partially similar action to be allowed for (Hewlett and Plackett, 1959). In order to do this we start from equations for the proportional non-response, Q, to the two drugs, since with independent action the general equation for P is less concise than that for Q. From the general equation for simple similar action, (*7.17*),

$$Q = \Pr\{w_1/\check{w}_1 + w_2/\check{w}_2 \leqslant 1\}. \quad (7.37)$$

Moreover from the opening statement of Section (7.10) we have for independent action

$$Q = \Pr\{w_1/\check{w}_1 \leqslant 1,\ w_2/\check{w}_2 \leqslant 1\}. \quad (7.38)$$

This equation states that Q is the probability that both inequalities within the brackets hold; (*7.34*), (*7.35*), and (*7.36*) are special cases of (*7.38*).

Introduction of a parameter, v, expressing the degree of similarity between the actions of the two drugs now leads to a general model for non-interactive action. We assume that v takes values of unity and zero for simple similar and independent actions respectively, and takes intermediate values for partially similar action. The most obvious general model is then

$$Q = \Pr\{vw_1/\check{w}_1 + w_2/\check{w}_2 \leqslant 1,\quad w_1/\check{w}_1 + vw_2/\check{w}_2 \leqslant 1\} \quad (0 \leqslant v \leqslant 1). \quad (7.39)$$

Putting $v = 1$ gives (*7.37*) and putting $v = 0$ gives (*7.38*). There are, in fact, an unlimited number of possible models giving the required extremes, and a simple alternative to (*7.39*) is

$$Q = \Pr\{(w_1/\check{w}_1)^{1/\xi} + (w_2/\check{w}_2)^{1/\xi} \leqslant 1\} \quad (0+ < \xi \leqslant 1). \quad (7.40)$$

Equation (*7.40*) sometimes has advantages over (*7.39*) in making the evaluation of Q easier. Although originally we arrived at (*7.39*) and (*7.40*) mathematically, we subsequently showed that (*7.39*) could result from a drug-receptor conception of partially similar action (Plackett and Hewlett, 1967).

Equations (*7.39*) and (*7.40*) are too general to allow direct evaluation of Q.

To do this additional assumptions have to be made, namely concerning the joint distribution of the tolerances $\check{w}_1$ and $\check{w}_2$. If probits are used a convenient assumption is that the joint distribution of log-tolerances is a bivariate normal. Making this assumption,

$$Q = \iint_R (2\pi)^{-1}(1-\rho^2)^{-\frac{1}{2}} \exp\{-(u_1^2 - 2\rho u_1 u_2 + u_2^2)/2(1-\rho^2)\} du_1 du_2 \tag{7.41}$$

Here ρ is the correlation coefficient. For (*7.39*), R is a region defined by

$$\begin{aligned} \text{hxp}\,[(Y-5+u_1)/\lambda_1] + \nu\,\text{hxp}\,[(Y_2-5+u_2)/\lambda_2] &\leqslant 1 \\ \nu\,\text{hxp}\,[(Y_1-5+u_2)/\lambda_1] + \text{hxp}\,[(Y_2-5+u_2)/\lambda_2] &\leqslant 1 \end{aligned} \tag{7.42}$$

For (*7.40*), R is a region defined by

$$\text{hxp}\,[(Y_1-5+u_1)/\xi\lambda_1] + \text{hxp}\,(Y_2-5+u_2)/\xi\lambda_2] \leqslant 1 \tag{7.43}$$

The fives in (*7.42*) and (*7.43*) appear because probits are used instead of normits (§ 3.2 and 3.7). Hewlett and Plackett (1959) give a fuller explanation of (*7.42*) and (*7.43*), and illustrate the boundaries of integration for certain numerical values of ν and ξ. They also illustrate some corresponding dose-response curves. When ν or ξ is unity and $\rho = +1$, (*7.41*) with (*7.42*) or (*7.43*) reduces to (*7.24*); when ν or ξ is zero, it reduces to (*7.34*) when $\rho = +1$, to (*7.35*) when $\rho = 0$, and to (*7.36*) when $\rho = -1$.

Hewlett and Plackett (1950) drew attention to the desirability of finding pairs of insecticides for which the tolerances are negatively correlated, because such pairs would reduce the chances of resistance appearing. Desultory efforts to find such pairs have since been made (see, for example, Brown, 1961), but without successful practical applications.

Although we shall not obtain models for interactive action as generalizations of (*7.41*), there is in principle no difficulty in doing so along lines indicated by Plackett and Hewlett (1952, Section 6).

7.12 Fitting to data of models for non-interactive action to drugs

The fitting to data of models for non-interactive action is not unduly difficult with electronic computers, and Ashford and Smith (1965) fitted some of their models using such computers. Not surprisingly, perhaps, the fitting of the very general model (*7.41*) has not been attempted. Without necessarily envisaging the use of computers, Hewlett and Plackett (1950) discussed the fitting of a general model for independent action; Mather (1940) examined the fitting of independent action with zero correlation of tolerances; and Plackett and Hewlett (1963) and Finney (1971) discussed methods for the fitting of models for simple similar action with complete positive correlation of tolerances.

The majority of work on the fitting of models of non-interactive action has taken place in the field of insecticides. Hewlett and Plackett (1950) found that data for the joint toxicity of certain pairs of insecticides agreed with the predictions of independent action, but the hypothesis was not convincing in view of disparate results from different techniques. Finney (1971) found that the model for independent action with zero correlation of tolerances accounted well for the joint action of digitalis and quinidine on frogs. Hewlett (1963c) found that mixtures of two insecticides, *n*-valone and dieldrin, were scarcely more toxic than the more toxic components, so that independent action with nearly complete positive correlation of tolerances was a possibility.

Turning now to simple similar action with complete positive correlation of tolerances, equation (*7.28*) accounted satisfactorily for a substantial amount of data (Plackett and Hewlett, 1963; Hewlett, 1963a, b and c). However, perhaps the most interesting successes were in the fitting of (*7.32*) to data for the joint action of two insecticides, methoxychlor and DDT (which are closely related chemically) (Plackett and Hewlett, 1963); and in fitting the appropriate extension of (*7.32*) to data for the joint action of four insecticidal pyrethroids (Sawicki *et al*, 1962). Here the probit-log-dose lines for the separate compounds in each investigation differed conspicuously in slope, so that under older ideas, a hypothesis of similar action would have been impossible. The extension of (*7.26*) also fitted the data for the pyrethroids. Equation (*7.31*) has been successfully fitted to a substantial number of sets of data (Hewlett, 1960; Finney, 1971; etc.).

Data for the joint toxicity of drugs may present difficulties in interpretation because more than one model will fit. It may then be helpful to adopt what might be called a 'network' approach. For example, suppose an attempt is to be made whether two drugs, A and B, are similar in action. Suppose that two other drugs, A′ and B′, known from other evidence to be very similar in action to A and B respectively, can be taken, and the joint actions of $A+B'$ and $B+B'$ are investigated. These joint actions should indicate which model can be expected to fit data for $A+B$ if these act similarly, and data for $A'+B'$ should provide further evidence. Hewlett (1963a, b, c) adopted an amplified approach of this kind.

7.13 Models from drug-receptor theory

Drug-receptor theory postulates that the action of a drug originates in chemical or quasi-chemical reactions between drug molecules and small specific areas of cells, known as receptors, in the tissues of an organism. Albert (1973) discusses the chemical and pharmacological aspects of the theory. It has been extensively utilized, among other things to account for the measured contractions of tissue strips, and the contractile effects of mixtures of drugs have been very usefully discussed by Ariëns *et al.* (1957) and by Ariëns and

Simonis (1961). Hewlett and Plackett (1964) suggested that with due caution models developed for the *in vitro* conditions could be extended to the more generalized quantal responses of whole organisms, e.g. paralysis and death.

Let us consider first a simple drug-receptor model, arising from the reaction of two drugs with the same set of receptors in the organism, such that the reactions are irreversible (e.g. because of covalent bond formation). Suppose that after reaction with drugs A and B a large excess of receptors remains unoccupied. Suppose that each unit of drug A acting results in its occupying n_1 receptors, so that w_1 of A acting will occupy $n_1 w_1$ receptors. Similarly w_2 of drug B acting will occupy $n_2 w_2$ receptors, and w_1 of A with w_2 of B will together occupy $(n_1 w_1 + n_2 w_2)$ receptors. Now suppose that more than N receptors must be occupied if quantal response is to occur. The condition for quantal response is therefore

$$n_1 w_1 + n_2 w_2 > N$$

that is

$$(n_1/N)w_1 + (n_2/N)w_2 > 1 \tag{7.44}$$

Putting $w_2 = 0$ gives the condition for response to drug A applied alone as $n_1 w_1/N > 1$. However, this condition is also $w_1 > \check{w}_1$, and hence $N/n_1 = \check{w}_1$, the action tolerance for A. Similarly $N/n_2 = \check{w}_2$. Relation (*7.44*) now becomes

$$w_1/\check{w}_1 + w_2/\check{w}_2 > 1$$

and

$$P = \Pr\{w_1/\check{w}_1 + w_2/\check{w}_2 > 1\} \tag{7.45}$$

which is (*7.17*), the basic equation for simple similar action. Derived models in terms of dose are the same as those following from (*7.17*).

Situations arising where there are reversible reactions between drugs and receptors have received more attention and are perhaps generally more interesting. Ariëns *et al.* (1957) have discussed a considerable number of models for graded response to combinations of drugs reacting with the same or different receptors. However, by applying the principle that a quantal response occurs when an underlying graded response exceeds some critical magnitude (Hewlett and Plackett, 1956; Ariëns and van Rossum, 1957), models for quantal responses can be derived from those for graded responses (see § 10.2).

We now proceed to one such derivation as an example. Hewlett and Plackett (1964) obtained models for quantal response under an assumption of competitive action, i.e. supposing that the two drugs react reversibly and competitively with the same set of receptors. Omitting several initial stages in the argument, we may state that physico-chemical considerations (Gaddum, 1937) combined with the concept of *intrinsic activities*, (Ariëns, 1954; Ariëns and Simonis, 1961) gives the graded response as

$$E(w_1, w_2) = (n_1 w_1 + n_2 w_2)/(1 + m_1 w_1 + m_2 w_2) \tag{7.46}$$

The quantal responses in competitive action can now be formulated by supposing that quantal response occurs when E exceeds some critical value, E_c. E_c is obtained by putting w_2 or $w_1 = 0$ and then substituting action tolerances $\check{w}_1$ or $\check{w}_2$ for w_1 or w_2 in (*7.46*). For drug A, say, applied alone, the graded response is

$$E(w_1, 0) = n_1 w_1/(1+m_1 w_1) \tag{7.47}$$

The critical graded response for drug A is

$$E(\check{w}_1, 0) = n_1 \check{w}_1/(1+m_1 \check{w}_1) \tag{7.48}$$

Thus the condition for quantal response to A alone in the individual organism,

$$E(w_1, 0) > E(\check{w}_1, 0) \tag{7.49}$$

reduces to

$$w_1 > \check{w}_1 \tag{7.50}$$

provided that $n_1 > 0$, a condition satisfied if A is an agonist, i.e. a drug active alone.

For the two drugs applied jointly there are three cases.

(*a*) *A is an agonist, B an antagonist* i.e. not by itself producing the quantal response. Here $n_2 = 0$, so that the graded response is

$$E = n_1 w_1/(1+m_1 w_1 + m_2 w_2) \tag{7.51}$$

Hence, using (*7.48*), quantal response occurs when

$$n_1 w_1/(1+m_1 w_1 + m_2 w_2) > n_1 \check{w}_1/(1+m_1 \check{w}_1) \tag{7.52}$$

i.e. when

$$w_1/(1+m_2 w_2) > \check{w}_1 \tag{7.53}$$

A comparison of (*7.14*), (*7.50*) and (*7.53*) shows that

$$Y = \zeta_1 + \lambda_1 \log\{w_1/(1+m_2 w_2)\} \tag{7.54}$$

Using (*7.12*), (*7.54*) becomes

$$Y = \alpha_1 + \beta_1 \log\{z_1/(1+\psi z_2{}^{\eta_2})\} \tag{7.55}$$

where $\psi = \mu_2 m_2$. Equation (*7.55*) is thus a model for the quantal response to an agonist, A, in the presence of an antagonist, B, and is particularly simple if η_2 is known or can reasonably be assumed to be unity. This model has not, so far as we know, been fitted to data.

(*b*) *Both A and B are agonists* i.e. each produce the response on its own. The maximum values of the graded responses given by the drugs separately are n_1/m_1 and n_2/m_2 respectively and we further assume these equal, i.e.

$$n_1/m_1 = n_2/m_2 \tag{7.56}$$

The critical graded response for A acting alone is given by (*7.48*) and similarly for B alone by $n_2\check{w}_2/(1+m_2\check{w}_2)$. If k' is the common value of these critical responses

$$\check{w}_1 = k'/(n_1 - m_1 k') \tag{7.57}$$

$$\check{w}_2 = k'/(n_2 - m_2 k'). \tag{7.58}$$

In view of (*7.56*) we obtain

$$n_1\check{w}_1 = n_2\check{w}_2$$

so that the distribution of either $\check{w}_1$ or $\check{w}_2$ determines the distribution of the other. The drugs together elicit the quantal response if

$$(n_1w_1 + n_2w_2)/(1 + m_1w_1 + m_2w_2) > k' \tag{7.59}$$

that is

$$w_1(n_1 - m_1k') + w_2(n_2 - m_2k') > 1$$

giving

$$P = \Pr\{w_1/\check{w}_1 + w_2/\check{w}_2 > 1\} \tag{7.60}$$

which is the same as (*7.17*) and (*7.45*). In view of the different derivations, the same equation can sometimes arise from non-interactive and interactive situations.

(*c*) A third case occurs when (*7.56*) is not satisfied, but, as Hewlett and Plackett (1964) pointed out, the resultant model is of doubtful practical value on account of the number of parameters involved.

7.14 Measurement of the potencies of drug mixtures

Although models for the quantal responses to mixtures of drugs have proved useful, for certain purposes less sophisticated approaches may suffice to provide adequate *measurements* of the toxicities of mixtures of drugs. Thus in experimental studies on synergists for insecticides, the LD50 of the insecticide alone and its LD50 in the presence of synergist, at a fixed ratio to insecticide, are often determined. The ratio of the first LD50 to the second is known as the synergistic ratio for the synergist concerned: this ratio is used to measure the efficiencies of synergists; the ratio of the synergist to the insecticide is fixed somewhat arbitrarily, partly in the light of the likely ratio for practical formulations (Hewlett, 1960).

In studying mixtures of insecticides, Hewlett (1963a, b and c) fitted (*7.29*) to his data and used estimates of λ as measurements of mixture potencies; the higher the potency the higher was the estimate of λ, which in principle was independent of the proportion at which a given pair of insecticides were included in the mixture. Sun and Johnson (1960) proposed what they called the *co-toxicity coefficient*. This is equal to 100 times the ratio of the observed toxicity of a mixture to its toxicity on the assumption of additive action. This coefficient has been used from time to time in work on insecticides.

For measurement of joint potencies of drug mixtures Hewlett (1969b) proposed what he termed the *joint action ratio*. This is applicable when there is synergism, potentiation or antagonism (or additive action) but not where there is coalitive action. Figures 7.3, 7.4, and 7.5 are isobolograms illustrating the derivation of the joint action ration, *R*. If only one drug, A, is separately

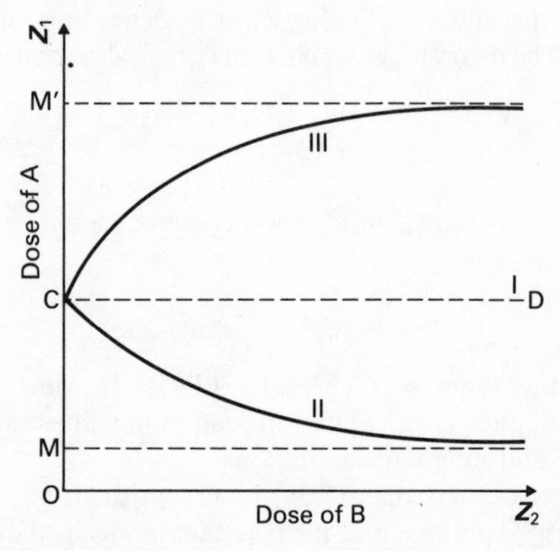

Fig. 7.3 Isoboles where one drug, B, is inert. The joint action ratio is OC/OM for synergism(II) and OC/OM′ for antagonism(III).

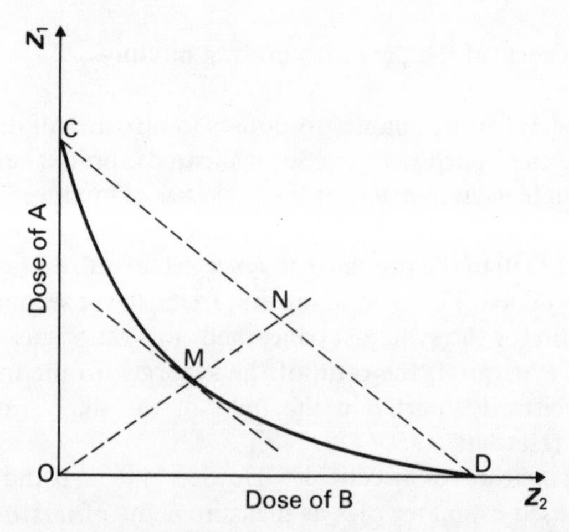

Fig. 7.4 Isobole for a pair of drugs separately active, showing potentiation. The joint action ratio is ON/OM.

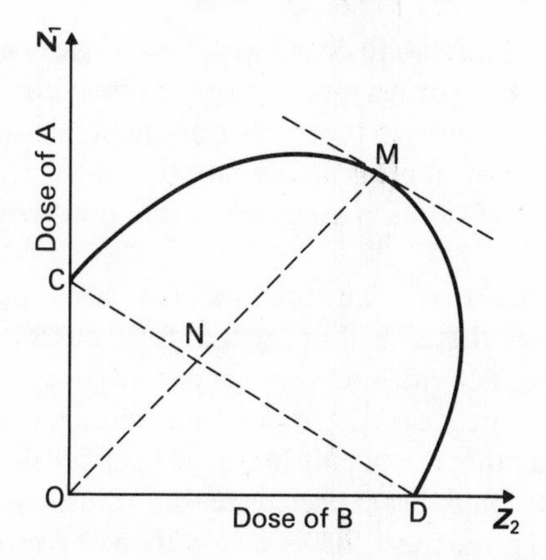

Fig. 7.5 Isobole for a pair of drugs separately active, showing antagonism. The joint action ratio is ON/OM.

active (Fig. 7.3), the isobole is typically a simple curve asymptotic to a line parallel to the line CD for addition. For synergism (isobole II), $R = \text{OC/OM}$, and for antagonism (isobole III), $R = \text{0C/OM}'$. If both drugs are separately active, a point M on the isobole furthest from the line for addition is taken; when potentiation occurs (Fig. 7.4), or antagonism (Fig. 7.5), $R = \text{ON/OM}$. Thus for synergism and potentiation R is the maximum ratio of the actual potency of the mixture of drugs to what their potency would be if their action was additive; for antagonism it is the corresponding minimum ratio. Whereas synergistic ratios and co-toxicity coefficients depend on arbitrary choices of doses, or ratios of doses, joint action ratios do not.

Hewlett (1969) showed how joint action ratios are related to mathematical models for joint action. For example, in two models for potentiation, (*7.4*) and (*7.5*), the parameters expressing the potentiation, η or κ, are related to R by

$$R = \tfrac{1}{2}(2^{1/\eta}) \tag{7.61}$$

$$R = 1 + \tfrac{1}{2}\kappa \tag{7.62}$$

He gave a number of examples, including graphical analysis, showing how R could be evaluated.

7.15 Design of experiments on joint drug action

The design of experiments for elucidating quantal responses to drug mixtures has been little investigated. However, the problem has been discussed in two publications. Mather (1940) in effect examined experimental design when testing whether response to a mixture exceeds that for independent action with zero correlation of tolerances, i.e. (*7.35*). He envisaged an experiment such that two groups of subjects were dosed with two single drugs respectively, and the third group with both drugs together in the same doses as the first two. He concluded that the responses to the single treatments should be no higher than 10%, that the number of subjects dosed with the single drugs should be equal, and that the number of subjects assigned to the joint treatment should be $1\frac{1}{2}$ times as great.

Finney (1971) discussed the planning of tests of (*7.31*), the most straightforward hypothesis among those for simple similar action. He envisaged that the activities of the separate drugs would be known approximately, in which case he considered that the mixtures should be chosen in such a way that the probit-log-dose lines for the separate drugs and for their mixtures could be expected to be about evenly spaced on the assumption of (*7.31*); the log-ED50's would then of course be evenly spaced. Plackett (unpublished) has discussed the choice between this strategy and an alternative in which the ED50's (instead of their logarithms) were evenly spaced.

8 Dose-response curves from heterogeneous populations and detection of resistance in insects

8.1 Dose-response curves

In bioassay work an experimenter normally standardizes the subjects as much as possible, controlling their age within limits and recording the results for each sex separately. Male and female animals often differ in their tolerance for drugs, but it is not always practicable to determine the sexes of animals such as insects in large numbers, so that batches of mixed insects may have to be used. The effect of this needs to be known. Moreover, in studies upon insects that are developing or have developed resistance to insecticides, genetically heterogeneous populations have often been encountered, containing, for example, a proportion of highly resistant individuals with the remainder normally susceptible. It is important to be able to interpret the atypical dose-response relations resulting, both in detecting resistance and in elucidating the genetics. The standard methods of fitting probit- and logit-log-dose lines are not applicable for the atypical curves, and graphical methods have been used in examining numerous sets of data. Hoskins (1960) and Tsukamoto (1963) have discussed the relations between probit response and log-dose of insecticide for heterogeneous populations encountered in studies on resistance.

Knowing the dose-response relations for two separate strains, the relation for a mixture of the two strains is easily obtained. Consider a certain dose of drug producing a proportional response, P_1, in a susceptible (S) strain, and a proportional response, P_2, in a resistant (R) strain. Suppose that a heterogeneous population, a mixture of the two strains, contains proportions m_1 and m_2 of S and R individuals respectively. Then the proportion of the heterogeneous population responding to the dose of drug in question will be

$$P = m_1 P_1 + m_2 P_2 \qquad (m_1 + m_2 = 1) \tag{8.1}$$

If one has a probit-log-dose line for each strain, then for any dose probit responses Y_1 and Y_2 can be found, and transformed to give P_1 and P_2. Using (*8.1*) gives P, which can be transformed to a probit, Y. Repeating the process for a number of doses gives a number of values of Y, so that a graph of Y against log-dose can be drawn.

First we consider the probit-log-dose curve for a heterogeneous population consisting of S and R strains, the former twice as susceptible as the latter, and each having a probit-log-dose line of slope 5. Male insects of a species are often of the order of twice as susceptible to insecticides as the females (Busvine, 1971). In the present example the probit-log-dose line for the S

Table 8.1 Calculations for plotting the probit-log-dose curve (H curve) for a heterogeneous population consisting of equal proportions of susceptible, S, and resistant, R, strains giving separate probit-log-dose lines separated by 1.5 probits.

S-line		R-line		H-line	
Y_1	P_1	Y_2	P_2	Y	P
8.0	.999	6.5	.933	6.82	.966
7.5	.994	6.0	.841	6.38	.917
7.0	.977	5.5	.691	6.01	.834
6.5	.933	5.0	.500	5.57	.716
6.0	.841	4.5	.309	5.19	.575
5.5	.691	4.0	.159	4.81	.425
5.0	.500	3.5	.067	4.43	.283
4.5	.309	3.0	.023	4.03	.166
4.0	.159	2.5	.006	3.61	.083
3.5	.067	2.0	.001	3.18	.034

strain is 1.5 probits above that for the R strain, as shown in Fig. 8.1, which also shows the curve, H, for a mixture of S and R in equal proportions. Table 8.1 gives the calculation of probits, Y, for plotting the H curve. Taking the second row of the table as an example, the dose for which $Y_1 = 7.5$ and $Y_2 = 6.0$ gives $P_1 = 0.994$ and $P_2 = 0.841$. $m_1 = m_2 = 0.5$ and equation (*8.1*) then gives $P = 0.917$ for which the probit, Y, is 6.38.

In Fig. 8.1 the H curve is almost straight in the range shown, departing by about 0.05 probit at most from exact linearity. Large experiments would be required to demonstrate the curvature of this H curve experimentally, though of course the average slope of the H curve is less than the slope of the S and R lines. The size of the probit difference, here 1.5, determines the departure of the H curve from linearity, so that this curve will be curved to the same extent for all pairs of S and R lines for which the product of the slope and the log-relative-potency is 1.5 (assuming of course $m_1 = m_2 = 0.5$). The foregoing example has been chosen to illustrate that a heterogeneous population can have two component strains differing appreciably in response, and yet give a probit-log-dose curve departing little from linearity.

Under some circumstances, however, heterogeneity in a population can be so pronounced as to produce conspicuous departures from linearity. This has occurred frequently in studies upon resistance of insects to insecticides (Georghiou, 1965; Brown, 1967; Brown and Pal, 1971). We thus take as a second example, a case in which the S and R strains differ more in susceptibility, giving probit-log-dose lines separated by 6.5 probits. If the slope of these lines is 5 as before, the R strain will be 20 times as resistant as the S. We again assume $m_1 = m_2 = 0.5$. Fig. 8.2 shows the markedly non-linear H curve

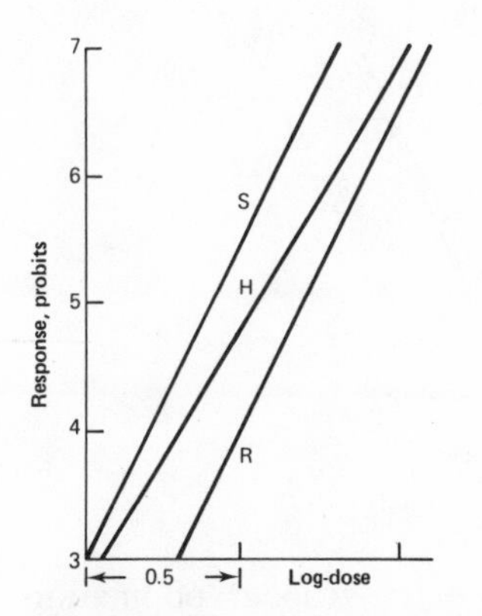

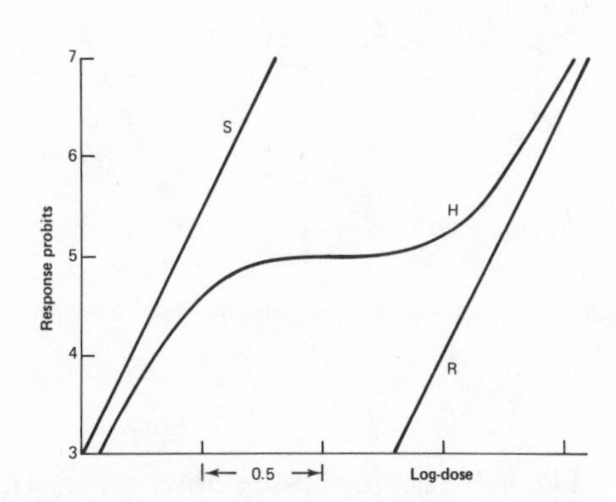

Fig. 8.1 The probit-log-dose curve, H, from a heterogeneous population consisting of a mixture of equal numbers of susceptible, S, and resistant, R, individuals giving straight probit-log-dose lines separated by 1.5 probits.
Fig. 8.2 The probit-log-dose curve, H, from a heterogeneous population consisting of a mixture of equal numbers of S and R individuals giving probit-log-dose lines separated by 6.5 probits.

for the population. Looking first at the lower portion of the S line, we see that Y_2 is so low that $P_2 \simeq 0$ and we can calculate $P \simeq 0.5\ P_1$ to a close approximation. For the upper portion of the H curve, near to the R line, Y_2 is so high that $P_1 \simeq 1$, and $m_1 P_1 \simeq 0.5$; thus we can here calculate

$$P \simeq 0.5 + 0.5P_2$$

to a close approximation. This middle portion of the H curve is nearly level, where $P_1 \simeq 1$ and $P_2 \simeq 0$, so that $P \simeq 0.5$ and $Y \simeq 5$. $Y = 5$ exactly at the point mid-way between the S and R lines, about which the H curve is symmetrical. The type of result illustrated in Fig. 8.2 can arise from a heterogeneous population of insects that is the progeny of a back-cross between heterozygous resistant individuals (genetically Rr) and homozygous susceptible (genetically rr), the gene for resistance being dominant (for examples see Brown, 1967).

The third and final example is similar to the previous, with $Y_1 - Y_2 = 6.5$ at each dose, but with $m_1 = 0.25$ and $m_2 = 0.75$; the individuals of the S strain are less numerous than those of the R. Fig. 8.3 shows the H curve for the population. For the lower part of the H curve,

$$P \simeq 0.25\ P_1$$

and for the upper part

$$P \simeq 0.25 + 0.75\ P_2$$

both to a close approximation. The curve is nearly level where $Y \simeq 4.33$, i.e. $P \simeq 0.25$. It sweeps nearer to the R line than to the S line. An H curve such as

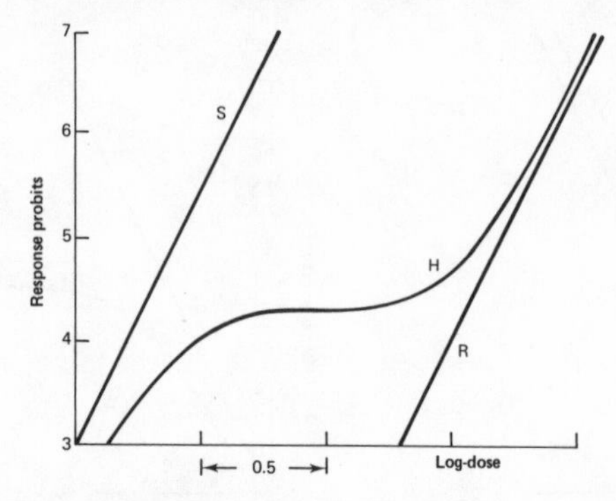

Fig. 8.3 The probit-log-dose curve, H, from a heterogeneous population consisting of one-quarter of S individuals and three-quarters of R individuals, the probit-log-dose lines separated by 6.5 probits.

this can be obtained if the population consists of the progeny of heterozygous individuals (Rr) (for example see Brown, 1967).

Many H curves of different forms have been calculated (see especially Georghiou, 1965; Tsukamoto, 1963). A number are more complicated than those illustrated here, especially if there are three or more strains mixed, but the principles of calculation remain the same. Logits can be used instead of probits, making little difference to the shapes of H curves. A dose, such as D indicated in Fig. 8.2, could be administered to a sample from a heterogeneous population in order to kill all individuals of one phenotype and leave those of another as survivors. Doses chosen on this principle for separating phenotypes are known as discriminating doses (Davidson, 1958).

8.2 Monitoring of insect populations for resistance

When an insecticide is applied at intervals for control of an insect pest, resistant strains may develop, necessitating a change of control measures. In order that a change can be made before serious failures in control occur, it is desirable to know when resistance is just beginning, and procedures have been developed for monitoring for resistance, especially in agricultural insect pests and those of public health importance (FAO, 1969; Brown and Pal, 1971; Champ and Campbell-Brown, 1970; Champ and Dyte, 1978). For a given pest species the general procedure is as follows: (1) An experimental technique for determining the response of the insects is adopted. (2) Before resistance has appeared, data for the relation between response and dose are obtained. These data are known as base-line data. (3) A dose is selected such as should give a very high response, e.g. 99.9 %, of susceptible insects. (4) Samples of insects are collected from the field at intervals, and the selected dose of insecticide is administered to them. The samples are often of 100 insects each, and in the absence of resistant individuals all will respond in most samples. If one or a very few fail to respond in, say, three successive samples further tests for resistance will be done with a series of different doses. If resistance is starting, the series of doses may give a probit-log-dose relation like the lower portion of the H curve in Fig. 8.2 or 8.3 (Brown and Pal, 1971). The selected high dose is known as a discriminating dose, the same term as mentioned in the previous section, but without quite the same meaning. The level of response required in the susceptible insects is often so high that the discriminating dose is fixed by extrapolation from the base-line data. This dose might differ a little according to whether probits or logits were used in analysis of the base-line data.

The aim in using discriminating doses will normally be to obtain complete response in most samples collected from the field before resistance occurs. It will therefore be of interest to know the probability of there being one or more survivors in a batch of normally susceptible insects treated with a discriminating dose. Suppose that the discriminating dose would cause a proportion P of an (infinitely large) batch to respond. The probability of it causing a complete response in a batch of n subjects is P^n, and that of one or more not responding is therefore

$$Q_n = 1 - P^n \qquad (8.2)$$

Table 8.2 The probability, Q_n, of one or more insects not responding in a batch of n (without any of a resistant strain) treated to cause a proportion P to respond in an indefinitely large batch.

	P				
n	0.95	0.98	0.99	0.999	0.9999
25	.7226	.3965	.2222	.0247	.0025
50	.9231	.6358	.3500	.0488	.0050
100	.9941	.8674	.6340	.0952	.0100
200	—	.9824	.8660	.1814	.0198
300	—	.9977	.9510	.2593	.0296
400	—	—	.9820	.3298	.0392
500	—	—	.9934	.3936	.0488

Table 8.2 gives values of Q_n for different values of P and n. As an example we see that the probability is 0.0952 of there being one or more individuals not responding in a batch of 100 treated with a dose equal to the ED99.9. The probability of there being just one individual not responding is lower than the corresponding value of Q_n, but it is the latter that is relevant in forming an impression of how a test may come out. The expectation of, for example, non-responders in five successive batches of 100 is the same as the probability for a single batch of 500. It should be stressed that Table 8.2 applies where no resistance has developed.

9 Time and response

9.1 Introduction

Experiments relating quantal response to time are less often done than those relating response to dose. However, the former are sometimes useful, especially if the interval between dosage and response decreases with increasing dose, so that the intervals are indicators of biological activity.

Time may enter in different ways into experiments in which quantal responses are observed. (1) Time of exposure to a drug may be used as a means for varying dose. For example, if insects are exposed to a given concentration of a fumigant, removed from the fumigant, and the response (usually death) determined after a suitable waiting period, then the response normally increases as the period of exposure to fumigant is longer. Often probit response is linear in log-period of exposure, and a probit plane often describes response as a function of both concentration and time (see § 3.8). (2) Time may be a time of response. If organisms of a batch are each given the same dose of drug, the time from dosage to response may be measured for each organism, or if all are dosed simultaneously the numbers that have responded can be observed at intervals. Response may represent a progression of drug action, such as paralysis, or a regression, such as recovery from an initial effect, e.g. anaesthesia, when the organisms are removed from the drug. (3) Time may act as both a means of varying dose and as a time of response, the two inextricably mixed, in the same experiment. Clearly this can happen if, say, aquatic organisms are immersed continuously in an aqueous solution of drug, and periods of immersion necessary for response are observed.

In this chapter we are concerned principally with times for response, though analysis of an experiment of the third type above will often be similar. We shall not be concerned with time as a means of varying dose.

Where time is time for response, an experiment tracing the rise with time of response in organisms given a fixed dose can be done in either of two ways. Firstly, if a series of batches of subjects are given the dose, the number responding in each can be determined, one from each batch, at respective different times. One count of responses is made on each batch. Probit or logit response could then be related to time suitably transformed. Secondly, numbers of subjects responding may be counted by a series of observations on each of one or more batches. This second procedure is more economical of experimental subjects, and is often justifiable, but the first may be necessary if (as can happen) the method of observation is such that the act of observation influences the occurrence of subsequent responses. Moreover, data collected by the second method will usually be easier to analyse, for, as will become clear in the next section, the response times are obtained as non-cumulative distributions.

The typical picture is that after dosage the response in a batch of subjects rises with time by a sigmoid curve, in due course reaching a maximum level of percentage though perhaps falling away later. Within a certain range of dose, the maximum level varies with dose, but at higher doses it is 100%. It is, perhaps, natural to expect that an individual subject will respond the sooner the more the dose of drugs exceeds its individual tolerance for the drug (see § 2.3). Often this is so, but not always (see Hewlett, 1974).

Analysis of time-response data usually includes attempts to describe time-response curves mathematically, and to estimate their parameters. The objects are similar to those in analysis of dose-mortality data in many ways, but the methods of estimation are usually different. Bliss (1937) made a basic contribution to the study of time-response data. Sampford (1952a) and Cox (1972) made further contributions. Bliss, and Sampford (1954) were much concerned with 'truncation', i.e. termination of observation before all responses had occurred. Hewlett (1974) examined some aspects of the dependence of dose-mortality relations upon time-mortality considerations. The studies just mentioned were on responses to single stimuli, especially single drugs, but Sampford (1952b) investigated responses to multiple stimuli, e.g. mixtures of drugs.

9.2 Analysis of a set of time-mortality data

A straightforward type of analysis is now described for a particular set of time-mortality data. This will illustrate how different analysis of such data can be from analysis of dose-mortality data.

Hewlett (1974) collected these data by observing daily the numbers of flour beetles dead in batches sprayed with solutions of DDT. Four different doses were used, but, although dose influenced the percentage dying, it did not influence the speed of lethal action, and nor did a beetle's sex. For these reasons, the results for different doses and sexes were pooled, giving the data recorded in Table 9.1, relating to a total of 204 deaths in a total of 547 beetles sprayed. There was no evidence that a count of deaths influenced subsequent ones. Hewlett (1974) included the results of an analysis of these data, but not the data themselves.

Owing to the multiple observations on each batch, we start from the numbers of beetles dying in the intervals between successive observations. The second column of Table 9.1 shows the numbers dying, f, 0–1 day, 1–2 days, and so on, after dosage. Obviously these numbers form an unimodal distribution, probably skewed to the right, so that the cumulative numbers plotted against days would fall on a sigmoid curve. However, it is convenient

Table 9.1 Times from dosage to death in beetles, *Tribolium castaneum*, treated with DDT. Analysis prior to class-mark computations.

Class limit days	No. dying in intervals f	Mortality at end of interval No.	%	Probits	Transf. class limit*	Class mark† x	Transf. class interval § i
0		0	0	—	0		
	9					0.893	1
1		9	4	3.25	1		
	42					1.069	0.119
2		51	25	4.33	1.119		
	54					1.161	0.077
3		105	51	5.03	1.196		
	50					1.227	0.057
4		155	76	5.71	1.253		
	19					1.277	0.047
5		174	85	6.04	1.300		
	11					1.320	0.039
6		185	91	6.34	1.339		
	8					1.357	0.035
7		193	95	6.64	1.374		
	3					1.389	0.030
8		196	96	6.75	1.404		
	4					1.418	0.027
9		200	98	7.05	1.431		
	4					1.443	0.024
10		204	100	—	1.455		

* Transformed class limit, i.e. limit in days to the power of 0.163.
† Mid-points of day classes to the power of 0.163, i.e. $0.5^{0.163}$, $1.5^{0.163}$, etc.
§ Differences between consecutive transformed class limits.

to plot them as percentages of the total of 204, as in Fig. 9.1. In Fig. 9.2, probits of these percentages are plotted against days. The latter plot is a test for normality, and the pronounced curvilinear trend shows that the distribution of the numbers dying is in fact skewed to the right.

A transformation of time such as normalizes the distribution enables the data to be described by meaningful parameters. A power transformation is an obvious candidate; if time is raised to a power, the transformation becomes stronger as the power decreases from unity, becoming a log-transformation in the limit as the power approaches zero, and becoming still stronger as the power becomes negative (see Finney, 1964, p. 66). Often a satisfactory transformation can be found by trial, for example by plotting the probits against square roots, logarithms or reciprocals of time. However, for the data of Table 9.1 a transformation of time to the power of 0.163 was arrived at by calculations that need not be given here. In Fig. 9.2 the probits have been

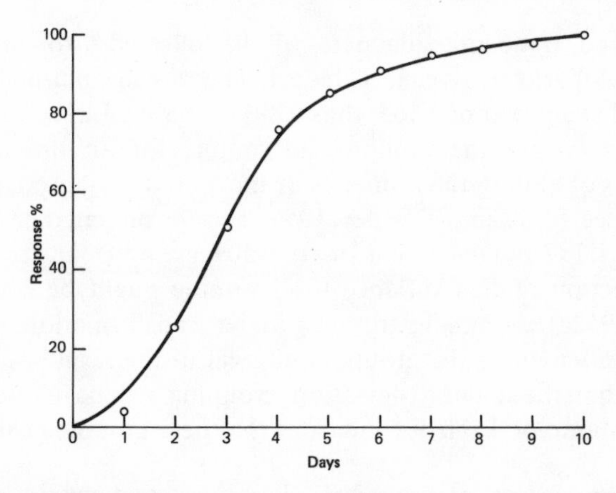

Fig. 9.1 Data of Table 9.1. Percentage of final number dead plotted against time from dosage in days.

plotted also against days raised to the power of 0.163, giving a reasonable fit to a straight line.

If one is used to dealing with dose-mortality data, the experimental points of Figs. 9.1 and 9.2 may seem to fit the curves surprisingly well. The reason is that here the ordinates of the different points are correlated with one another, a point discussed by Bliss (1937).

Having achieved a normalizing transformation, graphical estimates of median lethal time and the standard deviation of the transformed times may not be far wrong (see Table 9.2). However, calculation will usually be desirable; Sampford (1952a) gives a maximum likelihood method, but a

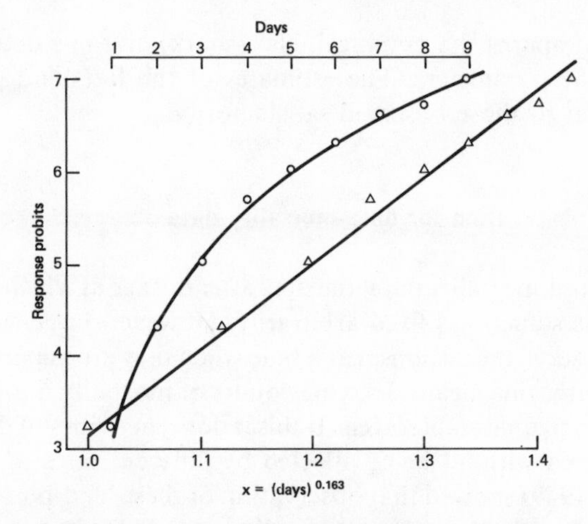

Fig. 9.2 Data of Table 9.1. Probits of the percentages in Fig. 9.1 plotted against time in days (circles) and against days raised to a power of 0.163 (triangles).

simpler method, frequently adequate, will be followed through here. This is based on class-marks, x, given in Table 9.1. They are the mid-points of the day classes raised to a power of 0.163; thus $0.893 = 0.5^{0.163}$, $1.069 = 1.5^{0.163}$, and so on. The calculations then follow the routine for an ordinary grouped distribution, computing the estimates of mean and variance from Sfx, Sfx^2 and $N = Sf$ (see, for example, Bailey, 1959). For the present data the mean was calculated as 1.187 and the variance as 0.01309, giving a standard deviation of 0.114. Correction of this variance for grouping might be considered, but Sampford (1952a), discussing grouping under transformation, suggests that, if the average length of the grouping interval in the neighbourhood of the mean is less than the standard deviation, grouping may be ignored. Reference to the last column of Table 9.1 indicates that here grouping can be ignored.

Table 9.2 Comparison of graphical, class-mark and maximum likelihood estimates of parameters of normal distributions fitted to response times raised to a power of 0.163 (data of Table 9.1).

		Graphical	Class-mark	Maximum likelihood
Mean		1.188	1.187	1.189
Standard deviation		0.083	0.114	0.109
Goodness of fit	X^2	—	—	5.1
	d.f.	—	—	4
	P	—	—	0.28

Table 9.2 compares the graphical, class-mark, and maximum likelihood estimates of the parameters. The estimates of the last kind give expected frequencies that fit those observed satisfactorily.

9.3 Time of observation for dose-mortality data

In obtaining dose-mortality data, the time after dosage at which the responses are observed is sometimes fixed arbitrarily. However, in general it appears desirable to observe the responses at a time when they are maximal. When the response is death, this means delaying counts of mortality until the numbers dead have risen to final stable levels. If this is done, activities of different drugs can be compared without being affected by different rates of action of the drugs. Beard (1949) showed that observation of these 'end-point' mortalities, which have a more-or-less absolute quality, avoided certain anomalies.

After investigating time-mortality relations in a particular experimental situation, Hewlett (1974) was able to elucidate the influence of time of observation on the dose-mortality relations. If the probit of end-point mortality was taken to be linear in log-dose, then probit mortality at earlier times of observation was theoretically curvilinear in log-dose. The curvilinearity could be obscured by scatter of experimental points, but the time-mortality relationships resulted in probit-log-dose lines for the earlier times being of lower slope than the line for end-point. Beard (1949) and Lloyd (1969) noted a similar influence on slope. Other things being equal, a higher slope permits a more precise estimate of log-ED50 than a lower one (see equation (*5.10*)), a further point in favour of end-point response.

10 The general nature of quantal responses

10.1 Introduction

In the first chapter of this book we outlined the concept of quantal responses, and in later chapters went on to discuss the interpretation of quantal responses in groups of organisms. However, we have not discussed the general nature of the quantal response of one individual organism to a drug. We now consider this, giving a shortened form of the account by Hewlett and Plackett (1956). One of the results is to show how dose-response relations for quantal responses are probably connected with the relations for graded responses, and hence to unify the mathematical representations of the two types of response.

We assume that the relation between per cent quantal response and dose of drug, if well controlled, represents a cumulative distribution of tolerances. In 1956, we discussed the validity of this concept, concluding that it is reasonable to attribute response of one individual and non-response of another to the same dose as due to an inherent difference between the two individuals during the period of action of the drug. Possible temporal variation in individual tolerance does not invalidate this idea. However, turning now to graded responses, it seems certain that in an individual organism, dose-dependent quantitative changes accompany the action of a drug, even if most of such changes are not so readily observable as the graded responses commonly used in bioassay. These changes might be alterations in the concentrations of certain substances within the organism (e.g. the accumulation of acetylcholine following administration of an anticholinesterase). Furthermore, it seems legitimate to regard any of such changes resulting from the drug action as graded responses.

10.2 A unifying hypothesis

We can now put forward a hypothesis that unifies the mathematical treatments of the two types of response; namely, *that an individual organism responds quantally if an underlying quantitative change, that is, a graded response, reaches a certain level of intensity characteristic of that individual organism.* If the dose of drug is insufficient to bring the quantitative change to the critical level, the quantal response will not occur. Hence the idea of a tolerance follows immediately. (For an application, see § 7.13.)

However, it should not be assumed that a quantal response in an individual organism can necessarily be related to *any* of the quantitative changes resulting from administration of the drug. Different quantal responses might stem from more or less distinct trains of events. If the situations for two trains could be represented thus

$$\text{Dose of drug} \begin{cases} \nearrow \text{quantitative changes A} \rightarrow \text{quantal response(s) A}' \\ \searrow \text{quantitative changes B} \rightarrow \text{quantal response(s) B}' \end{cases}$$

then A′ could most usefully be related to A and B′ to B. A′ could not usefully be related to B, nor B′ to A, unless the correlation between A and B happened to be very high.

10.3 Formulation of quantal responses

In the analysis of quantal data, we usually suppose that the tolerance distribution has a frequency function of the form $\beta f(\alpha + \beta x)$, where x measures the amount of drug on some convenient scale (very often logarithmic) and α, β are unknown parameters. The proportion of organisms responding to the drug is then

$$P = \int_{-\infty}^{\alpha + \beta x} f(t)\mathrm{d}t \qquad (10.1)$$

Normit analysis arises by taking

$$f(t) = \phi(t) = (2\pi)^{-\frac{1}{2}} \exp\left(-\tfrac{1}{2}t^2\right) \qquad (10.2)$$

and probit analysis similarly, with slight modification. Logit analysis arises when

$$f(t) = \lambda(t) \equiv \tfrac{1}{4} \cosh^2\left(\tfrac{1}{2}t\right) \qquad (10.3)$$

α and β are estimated by a process such as maximum likelihood.

10.4 Formulation of graded responses

On the other hand, when we are concerned with graded responses, each organism shows a response y, measured on a continuous scale, and thus provides more information than the binary classification of a quantal response. For each value of x, the responses are distributed among the population of organisms with a frequency function that can be converted to the normal, logistic, or any other form by suitably choosing the scale of y.

10.5 Connection of the two analyses

Following from Section 10.2, we connect the two forms of analysis in Sections 10.3 and 10.4 by postulating for each organism the existence of a *critical graded response*, of magnitude c, such that the quantal response occurs if the graded response exceeds c, but not otherwise. When x is given, the simplest possibility is that c has the same value κ for each organism, but more generally y and c will have some bivariate distribution among the population of organisms. To show how the tolerance distribution is derived from the graded response distribution, we consider two special cases.

(i) The critical graded response is a constant κ; and the distribution of graded responses has mean $(\theta+\psi x)$ and frequency function

$$f\{(y-\theta-\psi x)/\sigma\}/\sigma$$

where f is symmetrical about zero. If f is normal, σ is the standard deviation; if f is logistic, σ is (S.D.) $(\sqrt{3}/\pi)$. Then

$$P = \int_{\kappa}^{\infty} f\left(\frac{y-\theta-\psi x}{\sigma}\right)\frac{\mathrm{d}y}{\sigma} = \int_{-\infty}^{(\theta+\psi x-\kappa)/\sigma} f(t)\mathrm{d}t \qquad (10.4)$$

giving

$$\frac{\psi}{\sigma} f\left(\frac{\psi x+\theta-\kappa}{\sigma}\right) \qquad (10.5)$$

for the frequency function of the tolerance distribution. This formulation of P agrees with (*10.1*) provided that

$$\alpha = (\theta-\kappa)/\sigma \qquad (10.6)$$

and

$$\beta = \psi/\sigma \qquad (10.7)$$

(ii) The quantities y and c are jointly distributed with means $(\theta+\psi x)$ and κ respectively, in such a way that the distribution of $z = y-c$ has the frequency function

$$\frac{1}{\sigma} f\left(\frac{z-\theta-\psi x+\kappa}{\sigma}\right)$$

where f is symmetrical about zero. Then

$$P = \text{Prob. } (y-c>0) = \int_0^{\infty} f\left(\frac{z-\theta-\psi x+\kappa}{\sigma}\right)\frac{\mathrm{d}z}{\sigma} \qquad (10.8)$$

which again gives (*10.5*) as the frequency function of the tolerance distribution. Evidently the same tolerance distribution can arise from widely differing assumptions.

10.6 Testing the theory

Equation (*10.7*) is a very simple relation, and obviously experimental tests of this equation are possible. A test would consist of comparison of an estimate of β for a quantal response with one of ψ/σ for a graded response in the same organism; a valid comparison would require that the two responses resulted from the same train of biological events, as explained in Section 10.2.

Unfortunately comparisons of the kind just indicated have not yet been made. We can, however, cite indications in favour of (*10.7*) by quoting estimates of the two quantities, or rather their reciprocals, as shown in Table 10.1. There are obvious similarities both of position and spread in the distributions of the two quantities, lending indirect support to the theory.

Table 10.1 The frequencies of different values of estimates of $1/\beta$ for quantal responses and σ/ψ for graded responses, determined from a sample of bioassay methods for a variety of drugs, employing vertebrate material (values taken from Gaddum (1933) and Bliss and Cattell (1943)).

	Number of values	
Range of $1/\beta$ or σ/ψ	$1/\beta$ (quantal responses)	σ/ψ (graded responses)
0.0 –0.05	2	2
0.05–0.1	11	6
0.1 –0.2	23	13
0.2 –0.3	4	13
0.3 –0.4	4	8
0.4 –0.5	3	3
0.5 –0.6	1	0
0.6 –0.7	1	0
0.7 –0.8	2	0
0.8 –0.9	0	0
0.9 –1.0	1	0
Total	52	45

Appendix: Portable programmable calculators in logit and probit analysis

A.1 General

Section 6.5 referred briefly to the use of modern portable programmable electric calculators in logit and probit analysis. Further remarks on the topic are now made, although actual programs are not given, as these vary in detail according to the type of calculator (and the programmer).

The capacity of a calculator for doing programmed calculations depends on the number of its available program steps and the number of addressable (data) registers for storage of numbers. Both these numbers vary according to the type of calculator; moreover, in certain calculators the resources are variable, the number of steps being increased at the expense of the number of registers, or *vice versa*.

We outline here the use of the calculators for fitting logit – and probit-log-dose lines. They are readily programmed for fitting the former, and a program can easily be modified according to exact requirements. Writing more complex programs, such as for fitting parallel logit or probit lines, presents no special difficulties.

A.2 The logit-log-dose line fitted by maximum likelihood

As stated in Section 6.3, maximum likelihood fitting of a logit-log-dose line is an iterative process. In the fitting, two types of program can be used, which may be described as single-input and repeated-input. In the former the data are fed into the calculator once; in the latter they are fed in repeatedly, once at the beginning of each iteration (the latter procedure was alluded to in Section 6.5). The former is more convenient, requiring less effort from the operator, but requires more data registers and more program steps, i.e. a calculator of higher capacity.

Fig. A.1 is a flow diagram showing the general course of the single-input maximum likelihood fitting of a logit line. The initial data consist of four numbers for each dose level; the log-dose, x, the number of subjects, n, the expected response in logits, L, and the observed proportional response, p (corrected for control response). These data are put into data registers together with the control response, C, conveniently stored as $C/(1-C)$, and they remain there throughout. For k dose levels $(4k+1)$ registers are needed initially.

The initial data are indicated at the top of Fig. A.1. Processing starts at (1),

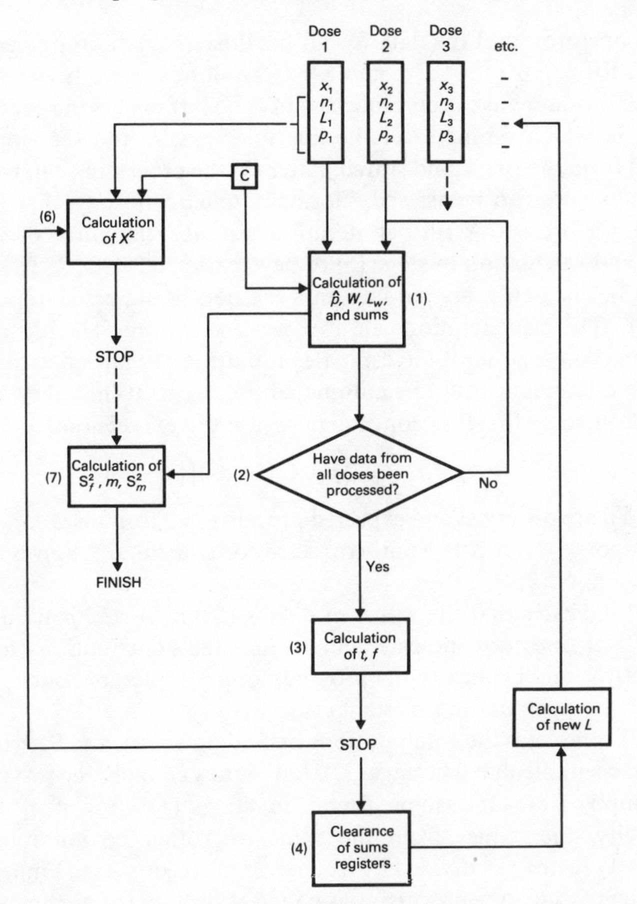

Fig. A.1 Flow diagram for single-input maximum likelihood fitting of a logit line.

in which the data for dose 1 are copied from the registers, together with the value of $C/(1-C)$, and the copies are used to calculate the following: (*i*) the expected proportional response, $\hat{p}$, from L; (*ii*) the weight of the observation, $W = nw$ (by (*6.3*)); and (*iii*) the working logit,

$$L_w = L + (p - \hat{p})/\hat{p}(1-\hat{p}) \tag{A.1}$$

Stage (1) continues with the computation of Wx, Wx^2, WL_w, and WL_wx, products that are summed into respective registers, as well as W. The conditional transfer (2) follows (1), (2) being a question in fact couched in numerical terms. A negative answer directs the processing back to the beginning of (1), so that the data for dose 2 are now processed, as the figure indicates. Stages (1) and (2), combined with the 'no' route, form a program loop that operates until the data for all the doses have been processed. The sums SW, SWx, SWx^2, SWL_w, and $SWL_w x$ will now have been stored.

After all the data have been processed in (1), (2) transfers the processing to stage (3), in which estimates of the line intercept, t, and the slope, f, are calculated from the sums, and stored. After (3) the processing can be stopped so that the values can be assssed. Normally one or more further iterations would be performed. A further iteration requires clearance of the sums registers, and calculation in stage (5) of new expected logits, L, from t and f found in (3). The new L are put into data registers as indicated, replacing the previous L. The next iteration can now proceed through (1), (2), and (3).

Iterations continue until the estimates t and f are regarded as sufficiently stable (the assessment could be automated if desired). When they are, X^2 is computed, in stage (6). It is convenient to use the expression

$$X^2 = S[n(p-\hat{p})^2/\{\hat{p}+C/(1-C)\}\{1-\hat{p}\}] \qquad (A.2)$$

Here p and $\hat{p}$ are observed and expected proportional responses corrected for control response, C. If P is $\hat{p}$ anti-corrected for control, (*6.9*) gives the same value for X^2 as (*A.2*).

Fig. A.1 indicates that the values of C, n, $\hat{p}$ (from L), and p are utilized in finding X^2. It does not indicate that in fact the contributions to X^2 are conveniently found by means of a program loop (simpler, of course, than the earlier one for computation of sums).

Stage (7) completes the computations by finding S^2_f, m, and $S_m{}^2$ (though m could have been calculated in stage (3)). If X^2 is low enough, the two variances follow simply from the sums found in stage (1); but if it indicates heterogeneity, the values from the sums will often be multiplied by a heterogeneity factor, X^2 divided by its degrees of freedom (see Finney, 1971). Thus the figure shows the processing as halted after stage (6), so that a decision can be made on whether a heterogeneity factor is needed.

The program just outlined requires at least 9 data registers in addition to the $(4k+1)$ needed for holding the data from an experiment of k dose levels (see above). Calling of the basic data from registers by 'indirect addressing' keeps down the total program steps required to about 300.

A.3 Shorter programs for logit analysis

With repeated input, shortened forms of the above program can be used for fitting of a logit line by maximum likelihood. With this form of input the basic data are in effect stored on paper, and program steps are not used in calling them.

For minimum chi-square fitting there is no iteration, so that storage of the basic data is superfluous. Moreover, only three numbers need to be fed in for each dose level, x, n, and p (or the corresponding empirical logit). The disadvantage of this form of fitting is that it does not deal adequately with zero and complete responses. Maximum likelihood fitting, on the other hand, does. However, the program described in the foregoing section will in fact do both forms of fitting; all that is necessary is to insert initially the empirical logit as that expected. The (initial) input is somewhat tautological, of course, but the first iteration is then a minimum chi-square fitting and subsequent ones maximum likelihood. In our experience this method normally gives a satisfactory maximum likelihood fitting in a total of 3 iterations. Incidentally, the fitting is completely objective, since the initial iteration does not depend on an eye-fitted line (a diagram is advisable, none the less).

A.4 Probit analysis

Maximum likelihood programs for fitting logit and probit lines are in most respects similar. However, each requires calculation of the proportion corresponding to the equivalent deviate (see stage (1) above), and here a difference arises. Calculation of the proportion corresponding to a logit requires 7 program steps and no datum register. On the other hand, ordinary calculation of the proportion corresponding to a probit requires at least 100 program steps and at least 3 data registers. However, with certain calculators the disparity is in effect less because they are provided with a special facility, namely of computing areas below the normal curve. This facility (which does not enlarge the main program), combined with about 40 additional program steps, enables a proportion to be found from a probit. Thus calculators of the type discussed here can be used for probit analysis, but this makes greater demands on program capacity than does logit analysis.

References

ALBERT, A. (1973). *Selective Toxicity*. 5th Edition. Chapman and Hall, London.

ARIËNS, E. J. (1954). Affinity and intrinsic activity in the theory of competitive inhibition. *Arch. Int. Pharmacodyn.*, **99**, 32–49.

ARIËNS, E. J. (1968). Introductory remarks. In: *Physico-chemical Aspects of Drug Action* (3rd International Pharmacological Meeting, Vol. 7). Ed. E. J. Ariëns, pp. 1–4.

ARIËNS, E. J. and VAN ROSSUM, J. M. (1957). Affinity, intrinsic activity, and the all-or-none response. *Arch. Int. Pharmacodyn.*, **113**, 89–100.

ARIËNS, E. J., VAN ROSSUM, J. M. and SIMONIS, A. M. (1957). Affinity, intrinsic activity and drug interaction. *Pharmacol. Rev.*, **9**, 218–36.

ARIËNS, E. J. and SIMONIS, A. M. (1961). Analysis of the action of drugs and drug combinations. In: *Quantitative Methods in Pharmacology* (Proceedings of a Symposium, Leyden, May, 1960). Ed. H. De Jonge, pp. 286–311. North-Holland Publishing Co., Amsterdam.

ARIËNS, E. J., SIMONIS, A. M. and OFFERMEIER, J. (1976). Interaction between substances in toxicology. In: *Introduction to General Toxicology*, pp. 155–71. Academic Press, London and New York.

ASHFORD, J. R. and SMITH, C. S. (1965). An alternative system for the classification of mathematical models for quantal responses to mixtures of drugs in biological assay. *Biometrics*, **21**, 181–8.

ASHTON, W. D. (1972). *The Logit Transformation (with Special Reference to its Uses in Bioassay)*. Griffin, London.

BAILEY, N. T. J. (1959). *Statistical Methods in Biology*. English Universities Press, London.

BEARD, R. L. (1949). Time of evaluation and the dosage-response curve. *J. econ. Ent.*, **42**, 579–85.

BERKSON, J. (1949). Minimum χ^2 and maximum likelihood solution in terms of a linear transform, with particular reference to bio-assay. *J. amer. statist. Ass.*, **44**, 273–8.

BERKSON, J. (1953). A statistically precise and relatively simple method of estimating the bio-assay with quantal response, based on the logistic function. *J. amer. statist. Ass.*, **48**, 565–99.

BLISS, C. I. (1937). The calculation of the time-mortality curve. *Ann. appl. Biol.*, **24**, 815–52.

BLISS, C. I. (1939). The toxicity of poisons applied jointly. *Ann. appl. Biol.*, **26**, 585–615.

BLISS, C. I. (1944). The U.S.P. collaborative cat assays for digitalis. *J. amer. pharm. Ass.*, **33**, 225–45.

BLISS, C. I. and CATTELL, MCK. (1943). Biological assay. *Ann. Rev. Physiol.*, **5**, 479–539.

BROWN, A. W. A. (1961). Negatively-correlated insecticides: a possible countermeasure for insecticide resistance. *Pest Control*, **29** (9), 24, 26, 40, 42, 44.

BROWN, A. W. A. (1967). Genetics of insecticide resistance in insect vectors. In:

Genetics of Insect Vectors of Disease. Eds. J. W. Wright and R. Pal, pp. 505–52. Elsevier, Amsterdam.

BROWN, A. W. A. and PAL, R. (1971). *Insecticide resistance in arthropods*. World Health Organization, Geneva.

BURNS, J. J., CUCINELL, S. A., KOSTER, R. and CONNEY, A. H. (1965). Application of drug metabolism to drug toxicity studies. *Ann. N. Y. Acad. Sci.*, **123**, 273–81.

BUSVINE, J. R. (1971). *A Critical Review of the Techniques for Testing Insecticides*. 2nd Edition, Commonwealth Institute of Entomology, London.

CHAMP, B. R. and CAMPBELL-BROWN, M. J. (1970). Insecticide resistance in Australian *Tribolium castaneum* (Herbst) – I. A test method for detecting insecticide resistance. *J. stored Prod. Res.*, **6**, 53–70.

CHAMP, B. R. and DYTE, C. E. (1978). *Report of the FAO Global Survey of Pesticide Susceptibility of Stored Grain Pests*. Food and Agriculture Organization, Rome.

COLQUHOUN, D. (1971). *Lectures on Biostatics*. Clarendon Press, Oxford.

COX, D. R. (1970). *Analysis of Binary Data*. Methuen, London.

COX, D. R. (1972). Regression models and life tables. *J. R. Statist. Soc. B*, **34**, 187–220.

DAVIDSON, G. (1958). Studies on insecticide resistance in anopheline mosquitoes. *Bull. Wld. Hlth. Org.*, **18**, 579–621.

DYTE, C. E. and ROWLANDS, D. G. (1968). The metabolism and synergism of malathion in resistant and susceptible strains of *Tribolium castaneum* (Herbst) (Coleoptera, Tenebrionidae). *J. stored Prod. Res.*, **4**, 157–73.

DYTE, C. E. and ROWLANDS, D. G. (1970). The effects of some insecticide synergists on the potency and metabolism of bromophos and fenitrothion in *Tribolium castaneum* (Herbst) (Coleoptera, Tenebrionidae). *J. stored Prod. Res.*, **6**, 1–18.

FAO (1969). Recommended methods for the detection and measurement of resistance. 1. General principles. *FAO Plant Prot. Bull.*, **17**, 76–82.

FINNEY, D. J. (1964). *Statistical Method in Biological Assay*. 2nd Edition. Griffin, London.

FINNEY, D. J. (1971). *Probit Analysis*. 3rd Edition. Cambridge University Press, Cambridge.

GADDUM, J. H. (1933). Reports on biological standards. III. Methods of biological assay depending on a quantal response. *Spec. Rep. Ser., Med. Res. Coun.*, No. 183. H.M. Stationery Office, London.

GADDUM, J. H. (1937). The quantitative effects of antagonistic drugs. *J. Physiol.*, **89**, 7P.

GEORGHIOU, G. P. (1965). Genetic studies on insecticide resistance. *Adv. Pest Control Res.*, **6**, 171–230.

GOLDSTEIN, A., ARONOW, L. and KALMAN, S. M. (1969). *Principles of Drug Action*. Harper and Row, New York and London.

GOODMAN, R. (1970). *Statistics*. 2nd Edition. English Universities Press, London.

HEWLETT, P. S. (1954). A micro-drop applicator and its use for the treatment of certain small insects with liquid insecticides. *Ann. appl. Biol.*, **41**, 45–64.

HEWLETT, P. S. (1960). Joint action in insecticides. *Adv. Pest Control Research*, **3**, 27–74.

HEWLETT, P. S. (1963a). Toxicological studies on a beetle, *Alphitobius laevigatus* (F). III. The joint action of doses of each of four toxicants put on two parts of the body. *Ann. appl. Biol.*, **52**, 305–11.

HEWLETT, P. S. (1963b). Toxicological studies on a beetle, *Alphitobius laevigatus* (F). IV. Joint-action experiments with two alkyl dinitrophenols. *Ann. appl. Biol.*, **52**, 313–9.

HEWLETT, P. S. (1963c). Toxicological studies on a beetle, *Alphitobius laevigatus* (F). V. The joint actions of some pairs of like and unlike toxicants. *Ann. appl. Biol.*, **52**, 351–9.

HEWLETT, P. S. (1968). Synergism and potentiation in insecticides. *Chemy. Ind.*, 1968, No. 22, 701–6.

HEWLETT, P. S. (1969a). The toxicity to *Tribolium castaneum* (Herbst) (Coleoptera, Tenebrionidae) of mixtures of pyrethrins and piperonyl butoxide: fitting a mathematical model. *J. stored Prod. Res.*, **5**, 1–9.

HEWLETT, P. S. (1969b). Measurement of the potencies of drug mixtures. *Biometrics*, **25**, 477–87.

HEWLETT, P. S. (1969c). The potentiation between thanite and arprocarb in their action on houseflies. *Ann. appl. Biol.*, **63**, 477–81.

HEWLETT, P. S. (1974). Time from dosage to death in beetles, *Tribolium castaneum*, treated with pyrethrins or DDT, and its bearing on dosage – mortality relations. *J. stored Prod. Res.*, **10**, 27–41.

HEWLETT, P. S. and PLACKETT, R. L. (1950). Statistical aspects of the independent joint action of poisons, particularly insecticides. II. Examination of data for agreement with the hypothesis. *Ann. appl. Biol.*, **37**, 527–52.

HEWLETT, P. S. and PLACKETT, R. L. (1956). The relation between quantal and graded responses to drugs. *Biometrics*, **12**, 72–8.

HEWLETT, P. S. and PLACKETT, R. L. (1959). A unified theory for quantal responses to mixtures of drugs: non-interactive action. *Biometrics*, **15**, 591–610.

HEWLETT, P. S. and PLACKETT, R. L. (1961). Models for quantal responses to mixtures of two drugs. In: *Quantitative Methods in Pharmacology (Proceedings of a Symposium, Leyden, May, 1960)*. Ed. H. De Jonge, pp. 328–36. North-Holland Publishing Co., Amsterdam.

HEWLETT, P. S. and PLACKETT, R. L. (1964). A unified theory for quantal responses to mixtures of drugs: competitive action. *Biometrics*, **20**, 566–75.

HEWLETT, P. S. and WILKINSON, C. F. (1967). Quantitative aspects of the synergism between carbaryl and some 1,3-benzodioxole (methylenedioxyphenyl) compounds in houseflies. *J. Sci. Fd Agric.*, **18**, 279–82.

HOSKINS, W. M. (1960). Use of the dosage-mortality curve in quantitative estimation of insecticide resistance. *Misc. Publ. ent. Soc. Amer.*, **2**, 85–91.

KUENEN, D. J. (1957). Time-mortality curves and Abbott's correction in experiments with insecticides. *Acta physiol. pharmac. nëerl.*, **6**, 179–96.

KUENEN, D. J., DEN BOER, P. J. and DE MELKER, J. (1957). Temperature influence on the mortality of *Calandra granaria* L. (Curcul. Coleopt.) from DDT. *Physiologia Comp. Oecol.*, **4**, 313–28.

LLOYD, C. J. (1969). Studies on the cross-tolerance to DDT – related compounds of a pyrethrin-related strain of *Sitophilus granarius* (L.). (Coleoptera, Curculionidae). *J. stored Prod. Res.*, **5**, 337–56.

LOEWE, S. (1953). The problem of synergism and antagonism of combined drugs. *Arzneimittel-Forsch*, **3**, 285–90.

LOEWE, S. (1959). Randbemerkung zur quantitativen Pharmakologie der Kombinationen. *Arzneimittel-Forsch.*, **9**, 449–56.

LOEWE, S. and MUISHNEK, H. (1926). Über Kombinationswirkungen. I. Hilfsmittel der Fragestellung. *Naunyn Schmiedebergs Arch. exp. Path. Pharmak.*, **114**, 313–26.

MAINLAND, D., HERRERA, L. and SUTCLIFFE, M. I. (1956). *Statistical Tables for use with Binomial Samples – Contingency Tests, Confidence Limits and Sample Size Estimates.* University College of Medicine, New York.

MATHER, K. (1940). The design and significance of synergic action tests. *J. Hyg., Camb.*, **40**, 513–31.

PEARSON, E. S. and HARTLEY, H. O. (1969). *Biometrika Tables for Statisticians* Vol. 1. 3rd Edition. Cambridge University Press, London.

PLACKETT, R. L. and HEWLETT, P. S. (1948). Statistical aspects of the independent joint action of poisons, particularly insecticides. I. The toxicity of a mixture of poisons. *Ann. appl. Biol.*, **35**, 347–58.

PLACKETT, R. L. and HEWLETT, P. S. (1952). Quantal responses to mixtures of poisons. *J. R. statist. Soc. B*, **14**, 141–63.

PLACKETT, R. L. and HEWLETT, P. S. (1963). A unified theory for quantal responses to mixtures of drugs: the fitting to data of certain models for two non-interactive drugs with complete positive correlation of tolerances. *Biometrics*, **19**, 517–31.

PLACKETT, R. L. and HEWLETT, P. S. (1967). A comparison of two approaches to the construction of models for quantal responses to mixtures of drugs. *Biometrics*, **23**, 27–44.

PURI, P. S. and SENTURIA, J. (1972). On the mathematical theory of quantal response assays. *Proc. 6th Berkeley Symp. Math. Statist. Prob.*, **4**, 231–47. Eds. L. M. Le Cam, J. Neyman, and E. L. Scott. University of California Press.

SAKAI, S. (1960). *Insect Toxicological Studies on the Joint Toxic Action of Insecticides.* Yashima Chemical Industry Co., Tokyo.

SAMPFORD, M. R. (1952a). The estimation of response-time distributions – I. Fundamental concepts and general methods. *Biometrics*, **8**, 13–32.

SAMPFORD, M. R. (1952b). The estimation of response-time distributions – II. Multi-stimulus distributions. *Biometrics*, **8**, 307–69.

SAMPFORD, M. R. (1954). The estimation of response-time distributions – III. Truncation and survival. *Biometrics*, **10**, 531–61.

SAWICKI, R. M., ELLIOTT, M., GOWER, J. C., SNAREY, M. and THAIN, E. M. (1962). Insecticidal activity of pyrethrum extract and its four insecticidal constituents against house flies. I. Preparation and relative toxicity of the pure constituents; statistical analysis of the action of mixtures of these compounds. *J. Sci. Fd. Agric.*, **13**, 172–85.

SNEDECOR, G. W. and COCHRAN, W. G. (1967). *Statistical Methods.* 6th Edition. Iowa State University Press, Ames.

SUN, Y-P and JOHNSON, E. R. (1960). Analysis of joint action of insecticides against house flies. *J. econ. Ent.*, **53**, 887–92.

TATTERSFIELD, F. and MORRIS, H. M. (1924). An apparatus for testing the toxic values of contact insecticides under controlled conditions. *Bull. ent. Res.*, **14**, 223–34.

TSUKAMOTO, M. (1963). The log dosage-probit curve in genetic researches of insect resistance to insecticides. *Botyu-Kagaku*, **28**, 91–8.

Index